Collaborative Cities

STEPHEN GOLDSMITH & KATE MARKIN COLEMAN

COLLABORATIVE CITIES

Mapping Solutions to *Wicked Problems*

Esri Press
REDLANDS | CALIFORNIA

Esri Press, 380 New York Street, Redlands, California 92373-8100

Printed in the United States of America
25 24 23 22 21   1 2 3 4 5 6 7 8 9 10

ISBN: 9781589485396

Library of Congress Control Number: 2021943859

For purchasing and distribution options (both domestic and international), please visit esripress.esri.com.

Contents

Foreword

When I was asked to review a book that combines maps and a practical understanding of local issues with a clear eye toward new and more collaborative models, I was curious. Well, after reading *Collaborative Cities*, consider me hooked.

Collaborative Cities comes at a critical juncture in America's history. At a time of deepening political divide and increasing national dysfunction, we cannot count on our federal government to solve all our problems. Yet, while just a third of Americans have any faith and confidence in the federal government, more than three-quarters of Americans still have great confidence in their local governments. Now, more than ever, it will be up to our communities and neighborhood leaders to address the many vexing, or "wicked," problems of our time. This book delivers a primer on place-based policy responses to these problems.

It has become cliché to say that cities are the future. But not all cities are the same. More people live in suburban and rural communities than in large urban centers. All these places, all these communities, all these neighborhoods, have distinctive assets, features, and character—as well as unique problems and challenges.

Legendary urban planner Jane Jacobs liked to say that a city is, after all, a federation of neighborhoods. As such, our country needs a local perspective to address inequality and rebuild communities. We face not just economic inequality but worsening spatial inequality. As the middle class and its once-robust neighborhoods have declined, our communities have fractured into areas of concentrated advantage and even larger areas of concentrated (often racially concentrated) disadvantage. Addressing this disparity requires place-based policies that can activate and strengthen local assets and the connective fiber between residents of disadvantaged neighborhoods and areas of economic growth and opportunity. The book's overarching perspective, its focus on cross-sector collaboration, and its use of maps and spatial analytics are all tools to empower and embolden neighborhoods and their residents to effect change.

Collaborative Cities puts a laser focus on the issues facing America's communities large and small, urban, suburban, and rural. The book tackles profound local issues including public health, sustainability, public safety, homelessness, social services, and environmental sustainability—all with an eye on how municipal and social sector leaders can work together to more effectively address these challenges.

The chapters of this book are filled with examples of tactical responses to these issues, from addressing child care and air quality to identifying ways to address inequity. More importantly, the book addresses how new and powerful tools of mapping and spatial analytics can help mayors, city governments, local stakeholders, nonprofits, and citizens join as partners to develop a broader, shared understanding of problems by presenting them in a foundation of real data. Take the growing issue of homelessness, for example: Effectively addressing the many facets of this issue requires understanding what, where, and how things are happening on the ground, who exactly is being displaced, and whether services are available to mitigate the challenges faced by people experiencing homelessness and where those services are located. Ideological posturing is of little help here. It's only through such clear-eyed, objective understanding that we can develop informed and shared narratives that will drive real change.

The key here is redefining leadership, not as the province of any one great figure, but as a process of collaboration—collaboration across political lines; across the private and public sectors; and across local government and its branches, the nonprofit community, labor, institutions of higher education, the medical community, and neighborhood and civic organizations. Collaboration means activating private and public organizations such as large corporations, foundations, universities, and medical centers that are literally anchored to their communities and whose activation has inspired such incredible change and revitalization in communities across America. Most of all, collaboration means forging new and more decentralized, bottom-up governance models that activate and engage the wider community.

Collaborative Cities is itself a collaborative effort between Stephen Goldsmith, former mayor of Indianapolis, and Kate Coleman, who led the YMCA through many of its most significant change initiatives. Goldsmith, a former deputy mayor of New York and now a Harvard professor, is widely known as one of the country's most effective and innovative mayors. Coleman served as executive vice president and chief strategist of the YMCA of the USA as a highly regarded local and national nonprofit sector leader. Their pairing may seem odd at first because they come from different sides of the political spectrum. But together, they provide a veritable guidebook for a pragmatic, nonideological, cross-sector approach to reflect the very best that America has to offer to address the wicked problems our cities face today.

It's clear that most Americans believe their local governments are up to the task. But those governments will need to build partnerships with stakeholders, workers, residents, and community leaders to get us on track and create the kind of places we are proud to call home. Coleman and Goldsmith point the way forward for all of us.

—Richard Florida

Richard Florida is one of the world's leading urbanists. Florida is a researcher serving as university professor at University of Toronto's School of Cities and Rotman School of Management and a distinguished fellow at New York University and Florida International University. He is a writer and journalist, having penned several global best sellers, including the award-winning The Rise of the Creative Class *and his most recent book,* The New Urban Crisis. *He has served as senior editor of* The Atlantic, *where he cofounded CityLab.*

Introduction

We are an unlikely pair: a liberal Democrat, an old-school Republican; a former social sector leader, a former mayor. We are divided in our politics, united in our civics. Over the years, we've learned how to find common ground, negotiate and share power, and build trust. In short, we've learned how to collaborate.

What's driving us? Mapping solutions to wicked problems.

Between us, we have decades invested in driving social outcomes. We've seen successes, and we've been frustrated by the intractability of so many of the complex, multidimensional problems deemed "wicked" more than 50 years ago by University of California Professors Horst Rittel and C. West Churchman. Examples of wicked problems include the pressing issues of homelessness, climate change, and childhood poverty. Based on our experience, and consistent with what is now a near constant refrain, we have long believed that "wicked problems," or "grand challenges," exceed the capacity of any one sector or set of actors. Instead, they demand the kind of creative thinking, democratized engagement, and integrated action that best happens across boundaries when government, nonprofits, businesses, and citizens work in concert.

As our friend and colleague Living Cities CEO Ben Hecht observed, "We can't 'nonprofit' our way out of our problems, nor can we fix them solely with government grants." And yet we are acutely aware how difficult it is to collaborate effectively across sectors. In this, we are not alone. Study after study begins with the observation that, at best, most multi-stakeholder efforts have mixed success. Ever optimistic, we believe the current decade has ushered in a Kuhnian moment (a moment named after American physicist and philosopher Thomas Kuhn) that involves fundamental changes in concepts and practices. Increasing emphasis on systems thinking, evolving ideas about organizational forms, and access to data and technology taken together change the calculus of cross-sector collaboration, which has motivated us to collaborate on this book.

Scores of articles and books have been published in the academic and popular press about cross-sector collaboration over the past several decades, especially since the turn of the century. Researchers—using organizational, institutional complexity, resource dependency, and communications theories (and the list goes on)—have provided important insights into the conditions that give rise to collaborative enterprises and the factors that influence their dynamics and contribute to their effective operation. As insightful as this work is, we think it would benefit from greater "how to" specificity, at least for the likes of us—government officials, nonprofit leaders, and citizens committed to creating social value. Technological developments in the areas of data, data analytics and visualization, connectivity, and the Internet of Things (IoT) have a significant role in facilitating the three distinct phases of collaborative efforts.

1. **Formation phase:** Bridge the perceptual and normative differences that inevitably exist among potential partners coming from different organizations or sectors, steeped as they are in their own beliefs about cause and effect. Mapping facilitates the community engagement and shared understanding that assists preliminary goal setting and is so critical in the formative stages of a collaboration.

2. **Operations phase:** Produce tangible outcomes in policy domains, where cross-sector actors join forces to address homelessness, health care, sustainability, and other wicked problems facing society.

3. **Adaptation phase:** Foster collaborative resiliency by helping partners iterate solutions and address new information and evolving tensions that often arise when discrete and often unequal partners come together attempting to accomplish a negotiated goal.

This book provides readers interested in multi-stakeholder action with a series of case studies drawn from our long experience working in the public, nonprofit, and private sectors. It presents a number of tools to improve collaborative processes and outcomes, including maps, GIS (geographic information systems), and analytics using information layered on maps. It is written by and for practitioners, and the case studies presented throughout are drawn from our lifelong experience across the public and nonprofit sectors.

Online resources

Throughout this book, we give dozens of examples of maps, apps, stories, documents, websites, and other resources supporting our premise about mapping and cross-sector collaboration. A companion website, CollaborativeCities.com, provides easy access to these resources. The companion website also contains information about other wicked problems such as coronavirus disease 2019 (COVID-19) and racial inequity—ongoing challenges that are rapidly evolving and could not be adequately addressed in the first edition of this book.

—Stephen Goldsmith and Kate Markin Coleman, August 2021

1

Why maps?

At its most basic, our work as government and community leaders is concerned with people and the factors that affect the quality of their lives. Which means that our work is about places—the places where people live, work, and play; where they experience problems; and where they craft and see solutions. Inadequate access to health care may be a complex problem affected by a combination of federal policy, the decline of blue-collar jobs, and medical liability claims. But if someone has no health insurance and gets sick, they experience it in their neighborhood, their home, their bed. What is rooted in systems is experienced in place, which is why place-based thinking is so important to the design and delivery of social services. It is also why geocoded data can serve as a powerful tool in the collaborative toolbox. This book shows that data collected, analyzed, and visualized geographically has a unique and powerful role in the formation and operation of collaborative enterprises and in their ability to adapt their understanding of, and response to, complex social issues.

Our experience animates our interest in geospatial matters. One of us, Kate, served as executive vice president for branding and strategy at the YMCA of the USA, a federated organization of more than 20 million members. At headquarters, problems looked national, but each local affiliate was separately incorporated, with deep community roots, a local board and local fundraising, and services configured for its community. Of course, a creative tension existed between national policy and goals and local implementation. Even a national strategic plan was achieved and configured locally because implementation and need were contingent on place.

In the public sector, Stephen, in his work as a mayor of Indianapolis and as deputy mayor of New York, has been constantly struck by how much localized tacit knowledge doesn't make it to city hall in the implementation of programs. His plans in structuring community interventions involving young mothers looking for work and childcare frequently involved searching for the right local community-based and faith-based organizations and volunteers with which government could partner. As mayor and deputy mayor, most of his policy development and operations experience involved organizing disparate community groups and city agencies into place-based strategies.

Maps: A common platform for understanding

Maps have two critical properties that make them particularly well-suited to the demands of cross-sector collaboration. First, multiple sets of data can be combined or layered onto maps in ways that would "otherwise be difficult to visualize and analyze."[1] To paraphrase James Minton, who works with the Eviction Lab at Princeton Univeristy, unlike a graph or table, you can saturate a map with information without creating cognitive overload. Beyond the inevitable insights that come from location intelligence, why is this important? Because it creates a common reference point for potential partners to translate their distinct world views into shared understanding. In the formative stages of collaboration, shared understanding has proven to be an important factor driving later success. Once formed, maps provide a platform upon which emerging and experiential data can be layered, creating a base of new, democratized knowledge that allows partners to adapt their interventions in something approaching real time.

The second property of maps that makes them such an important tool for collaboration is less tangible but no less important. Humans by their very nature situate themselves in place in an absolute sense and in relation to others.[2] "Me" on a map becomes "me" in relation to the data layered on the map. "Me" becomes "we," and "we" signals interdependence. In addition to shared understanding, recognition of interdependence is an important factor driving interest in joining together. It is also important in the development of collective identity and, ultimately, collective agency. If we believe our neighbors will join with us and the local community organization to clean up the vacant lot or spruce up the flowers and grass along the roads, we will be much more likely to take collective action. These acts in turn increase our collective agency and enhance the resulting social capital—the network of relationships among people who live and work in a particular society—that improves our neighborhood.

A new model for cross-sector collaboration

This book adopts a broad definition of *collaboration* from academic experts Kirk Emerson, Tina Nabatchi, and Stephen B. Balogh, as "the processes and structures of public policy decision-making and management that engage people constructively across the boundaries of public agencies, levels of government, and/or public, private, and civic spheres in order to carry out a public purpose that could not otherwise be accomplished." This definition acknowledges the dynamic state of the field. New forms of collaborative enterprises arise constantly. They include everything from "formal governing arrangements...to hybrid arrangements such as public-private and private-social arrangements and co-management regimes... to intergovernmental collaborative structures...to civic engagement."[3]

These varied arrangements can be short-lived or transactional in nature, as when residents partner with a community-based organization and municipal government to turn a vacant lot into an urban park. Some are of longer or even indeterminate duration, frequently taking the form of networks that link existing organizations in a joint effort to address a specific but multifaceted social issue. For instance, Chicago is home to numerous community-centered networks, but their action taken to "reduce violence, improve schools, [and] develop affordable housing" occurs in specific places and mostly involves a range of community and public resources. As the authors of the report Network Effectiveness in Neighborhood Collaborations note, this is to be expected in a city with a long history of operating based on connections.[4]

Using social network analysis, the report examines how certain partnership patterns support better collaborations than other attempts at collective work.

The emerging interest in collective impact, which brings many parties together for a common objective, highlights the needs for spatial visualization tools. The most profiled example of a collective impact initiative, Strive was born in Cincinnati and Northern Kentucky in 2006 and focused on improving educational outcomes.[5] It serves as the model for hundreds of such partnerships around the world.[6] In addition to transactional and networked arrangements, partnerships that have roots in the collaborative governance itself sometimes produce an institutionalized framework with an ongoing charter and steady funding stream.

Irrespective of their duration, most cross-sector collaborations engage in a series of activities in three phases:

1. Formation
2. Operations
3. Adaptation

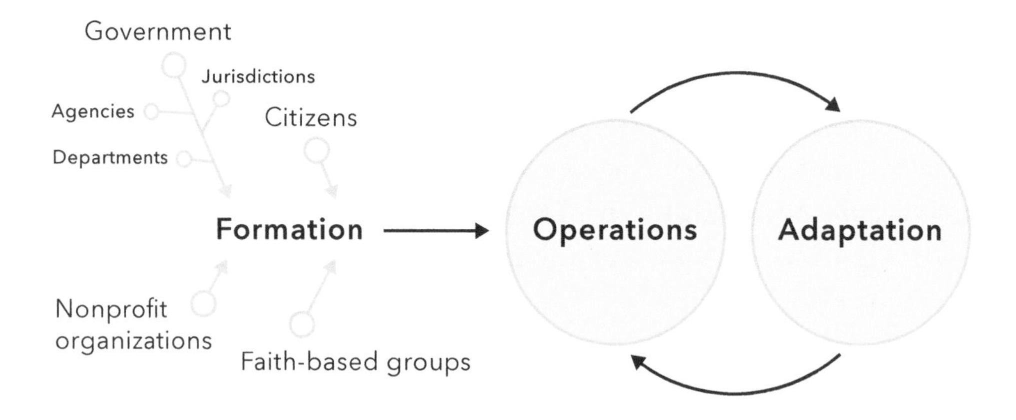

Figure 1.1. Workflow for the formation phase of cross-sector collaborations.

Formation: Creating a shared understanding

Most researchers note that collaborative enterprises emerge under a set of conditions that affect the quality of their formation. These antecedent conditions[7] may be political (such as mandates from the US Department of Housing and Urban Development), economic (a dedicated funding stream), or situational (a history of sector failure). These conditions constitute the larger context in which the collaboration operates and in addition to giving rise to it, they impact its ongoing dynamics.

However, even in the face of favorable conditions, an effective collaboration typically takes the presence of some number of drivers to catalyze its formation.[8] One of these drivers is a leader who can frame the problem to be addressed in a way that is compelling and helps potential partners see the stake they have in working together to resolve it. A nonprofit or governmental leader, for example, can highlight locations where drug overdoses are most severe and increasing and can call for a comprehensive, multiparty effort, resulting in the reallocation of resources to solve the problem. Community activists also can drive the formation of a cross-sector collaboration by mapping and locating a problem—for example, income and service inequality tied to few jobs and poor transportation.

Understandably, potential partners come to the table with their own particular, organizational-bound views of reality. Their beliefs reflect their assumptions about the factors that give rise to problems and the most effective practices in addressing them. These beliefs, premised on their experience and the information to which they have access, form an incomplete picture of reality. For a collaborative to progress beyond an initial convening,[9] potential partners must reach a shared or negotiated understanding of the issue. This understanding includes its "symptoms, causes, [and] assumptions [regarding]...potential solutions."[10] This critical first step allows the group to "form mental models...and negotiate the terms upon which to come together,"[11] including typically their preliminary aims, intended impacts, and theory of action. These provisional agreements set the early direction of the collaboration, shaping its structure and the norms that guide how partners work together.

The process of developing a shared understanding or definition of the problem is enhanced using sense-making approaches—processes that help make sense of an ambiguous situation by creating situational awareness. Groups arguing about the effect of a proposed building on their neighborhood bring differing contexts. For example, the developer cares about profitability, the economic development director cares about jobs, and local community leaders care about quality of life in terms of traffic and shadows. Augmented reality mapping tools can now at least let people who want to collaborate but disagree go beyond individual imaginations to demonstrate the effect of real decisions.

Operations: Cross-sector tasking in action

Cross-sector collaborations form to address a range of social issues with varied intended impacts as their objectives. These collaborations are particularly apt where underlying constitutive issues interact in complex and difficult-to-parse ways, as in the case of wicked problems. Their aims cover a spectrum of intended impacts, from improving the educational outcomes of children living in a given school district to improving air quality to reduce the numbers of asthmatic children and the amount of Medicaid expenses. The tasks have differing input costs and different outcomes, about which reasonable people often disagree. Coordinating and integrating the delivery of service among like and complementary providers, disbursing resources based on community input, and increasing education and awareness all depend on shared understandings of the relationship of the problems and interventions. This book highlights examples of cross-sector collaborations formed to address the wicked problems of climate change, homelessness, equity, access to health care, and delivering emergency services. Not surprisingly, the factors giving rise to the need for such collaborations constitute a web of interrelated issues with a long history of suboptimal government and social sector intervention.

Naturally, the task environment or specific operations of the collaborative depend on its aims and theory of action. The next chapters profile how collaborations use location intelligence (open and proprietary geocoded data, visualizations, and analytics) to improve their operations, especially in making trade-offs and choices about the best or most desirable interventions and outcomes. Mapping health and emergency problems affects the speed of response and can lead to community-based interventions that reduce the problems before the 911 call, such as for homelessness or opioid overdoses. The changing location of open jobs and underemployed individuals allows neighborhood groups, transit officials, and employers to craft transportation solutions that will need constant calibration. Together with social media, parents coming together with school and safety officials can map safe walking paths. Open data visualizations that allow comparisons of problems and solutions can assist one neighborhood in its advocacy for fairer distribution of city services.

In other words, GIS visualization will allow the collaboratives to more quickly and fully determine the array and quantity of tasks. This spatial analysis of needs and assets will affect the formation and the operation of the cross-sector collaborations. It may identify a need that requires the participation of a new service provider partner—for example, the need to address domestic violence as part of the homelessness cross-sector collaboration. Or it may require a new responsibility by an existing participant. Perhaps the school social worker, already a member through the school, refers vulnerable children and their families to an organization that can provide housing assistance.

Adaptation: Iterating toward better interventions

Once formed and underway, successful cross-sector collaborations can incorporate an array of information and perspectives into their approach and simultaneously develop a sense of collective agency. It is in the process of negotiating and iterating solutions that they derive their unique power and ability to innovate synergistic action, and thus they are able to bridge differences, build trust, and further their "capacity for joint action." This is particularly critical in the domain of wicked problems.[12]

The word *wicked* is used to describe societal problems such as homelessness, access to health care, and climate change not because they are in any way immoral but because they feature "multiple, overlapping, interconnected subsets of problems that cut across policy domains."[13] Because they are "endemic, multi-scalar, and evolving,"[14] they are not subject to any sort of "stopping rule."[15] Which means they are most effectively addressed in a dynamic fashion, with new knowledge continuously incorporated into the design and delivery of collaborative actions. Collaborations cannot solve these problems but can influence them so they shrink or change. Mapping the growth or abatement of a social problem—for example, the numbers and locations of people experiencing homelessness and information about their changing health care needs—should help the collaboration adapt to meet those needs. If information is presented clearly and contextually, open data, Internet of Things (IoT) sensors, and mobile devices allow for frequent and adaptive iterations.

The role of maps

With this additional information in mind, we return to the role of maps in the three phases of cross-sector collaboration—formation, operations, and adaptation. Maps tell stories about places and the people who live there. The value of maps lies partly in the conversations they generate when partners use maps to negotiate the definition of need, risk, response, and impact. This negotiation binds partners together and creates a foundation from which collaborations can build a coordinated response, allocate resources, and iterate solutions. When multiple parties agree on the meaning of neighborhood data, trends, or impact evaluations, they come to share assumptions, norms, and ways of operating. Successful negotiation creates unity. "Me" becomes "we."

Real-time geotagged data can also uncover an unintended consequence of the collective action, allowing adaptation with less acrimony. The process of map-aided discussions of how to confront a problem plays a critical part in the formation of the cross-sector collaborations and in their subsequent operation and adaptation. Let's take an example referenced later in the book about how to involve the community in the design of a neighborhood park. Conversations aided by visualizations show numbers and age of users and what their activities are in the park. The discourse about that information would be enriched through transportation and trail information reflecting perhaps that a group of potential users cannot get to the park, spurring a search for solutions. Even after cross-sector groups reach an agreement on what amenities the city will build in the park and which volunteer groups will provide programs, the amount of available and continuing data might show unanticipated use patterns. Presenting the data visually aids the discourse that facilitates iterations of the original plans and, in turn, creates collective agency—the city's park becomes "our" park, an expression of our collective agency.

How this book is structured

The book lays out a path for producing more public value through collaborative action made possible through location intelligence. Each chapter will address the formation, operations, and adaptation of cross-sector collaborations but with attention to a specific domain.

Chapter 2 examines civic engagement and how mapping plays a critical role in mobilizing public action. Sometimes, a public official creates a map to convince the residents to collaborate and take collective action in addressing an important issue. But other times, community groups might use the mapped data to draw attention to an important problem. Maps also can aid in the translation of civic advocacy into action—from engagement of one's voice to engagement of one's actions.

The next five chapters address wicked problems in specific policy domains: chapter 3, social services; chapter 4, public health; chapter 5, homelessness; chapter 6, public safety; and chapter 7, sustainability. Each chapter examines how community, nonprofit, for-profit, and government leaders respond to important problems through cross-sector collaborations facilitated by spatial analytics. By domain, we describe how location intelligence is being used to design and deliver service, allocate resources, engage the community, and otherwise facilitate the achievement of the collaborative's goals. Chapter 8 delivers new hope for cross-sector collaboration.

These chapters demonstrate that, with technological advances, the operation of cross-sector efforts has a newly enabled opportunity to produce dramatic change. New digital tools, real-time IoT data, handheld mobile tools for workers and residents alike, and cloud-based mapping that easily allows the display and analysis of data can vastly improve insight, trust, and the efficacy of collaborations. Despite the seriousness of the wicked problems facing cities around the world and the serious differences of opinion within the public and certainly among politicians, a shared understanding can lead to effective action and better communities.

Notes

1. Peter Folger, "Geospatial Information and Geographic Information Systems (GIS): Current Issues and Future Challenges," *Congressional Research Service,* June 8, 2009, https://fas.org/sgp/crs/misc/R40625.pdf.

2. Luke Whaley, "Geographies of the Self: Space, Place, and Scale Revisited," *Human Arenas* 1, no. 1, March 2018: 21–36, https://doi.org/10.1007/s42087-018-0006-x.

3. Kirk Emerson, Tina Nabatchi, and Stephen Balogh, "An Integrative Framework for Collaborative Governance," *Journal of Public Administration Research and Theory* 22, no. 1, 2011: 1-29, https://doi.org/10.1093/jopart/mur011.

4. David M Greenberg and others, "Network Effectiveness in Neighborhood Collaborations: Learning from the Chicago Community Networks Study," *Manpower Demonstration Research Corporation,* November 2017, https://www.mdrc.org/sites/default/files/CCN_Report_ES_Final-web.pdf.

5. Jennifer Splansky Juster and Sheri Brady, " Launching a New Study on the Results and Lessons Learned from Collective Impact Initiatives," *Collective Impact Forum,* April 6, 2017, https://collectiveimpactforum.org/blogs/700/launching-new-study-results-and-lessons-learned-collective-impact-initiatives.

6. Ibid.

7. John M. Bryson, Barbara C. Crosby, and Melissa Middleton Stone, "The Design and Implementation of Cross-Sector Collaborations: Propositions from the Literature," *Public Administration Review,* 66, 2006: 44–55, www.jstor.org/stable/4096569.

8. John M. Bryson, Barbara C. Crosby, and Melissa Middleton Stone, "Designing and Implementing Cross-Sector Collaborations: Needed and Challenging," *Public Administration Review*, 75, August 9, 2015: 647-663, https://doi.org/10.1111/puar.12432.

9. Barbara Gray and Jill Purdy, *Collaborating for Our Future: Multistakeholder Partnerships for Solving Complex Problems* (New York, NY: Oxford University Press, 2018).

10. Cynthia Hardy, Thomas B. Lawrence, and David Grant, "Discourse and Collaboration: The Role of Conversations and Collective Identity," *The Academy of Management Review* 30, no. 1, 2005: 58–77, www.jstor.org/stable/20159095.

11. John W. Selsky and Barbara Parker, "Cross-Sector Partnerships to Address Social Issues: Challenges to Theory and Practice," *Journal of Management* 31, no. 6, December 2005: 849–873, https://doi.org/10.1177/0149206305279601.

12. Hardy, 58–77.

13. Horst W. J. Rittel and Melvin M. Webber, "Dilemmas in a General Theory of Planning," *Policy Sciences* 4, no. 2, June 1973: 155.

14. Selsky, 849–873.

15. Horst W. J. Rittel, and Melvin M. Webber, "Dilemmas in a General Theory of Planning," *Policy Sciences* 4, no. 2, 1973: 155–169, https://www.jstor.org/stable/4531523.

2

Mapping civic engagement

This book presents a straightforward proposition: Through maps, effective city leaders can tell stories that bring together individuals from the nonprofit, government, philanthropic, and business sectors and frame a narrative that prompts collective community action.

As mayor of Indianapolis from 1992 to 2000, co-author Stephen Goldsmith planned a series of controversial moves that involved using private-sector companies to bring management efficiencies to a range of city services. For example, by using an international water company to manage the wastewater system and working with the public labor union, the city could free up hundreds of millions of dollars for community investments. Yet these changes in the management of a public-private partnership by themselves did not generate enough support for council approval. Goldsmith needed to frame the project narrative around the benefits from the new management approach. The Building Better Neighborhoods campaign mapped each capital project that would be finished using the savings. To catalyze broad community action, the campaign placed blue-and-white signs in front of each affected sidewalk, police station, library, or park.

The strength of a narrative in these scenarios—the ability to inspire collaborative action—is best reinforced when maps tell stories. This book features examples that demonstrate the power of framing problems or initiatives through a spatial lens to support multi-stakeholder and cross-sector networks. As the collaboration empowers individuals, civic groups, and city planners to work together, stimulating civic engagement, the resulting social capital strengthens the entire community. That engagement, in turn, creates locational intelligence that furthers shared goals.

In one example, discussed in chapter 4, health care professionals in Klamath County, Oregon, brought government and community players together in a broad effort to improve health. Another example, in chapter 7, shows how public officials inspired a range of governmental and volunteer organizations to protect Mississippi River watersheds. In both cases, the collaborating entities shared goals, information, maps, and other resources as part of an organized effort.

A unique property of maps is that they enable individuals to situate themselves in relationship to others. Spatial awareness helps them see that they share a common identity or challenge. The senior official who wishes to rally different stakeholders—whether members of an organization or voters of a city—must present wicked problems, such as homelessness or opioid addiction, in an informed and contextualized way. Visualizing questions and answers in a layered fashion enhances that conversation.

Maps encourage civic engagement because they help inform residents of neighborhood conditions, enable officials to track changes in those conditions over time, locate important community assets, and bring parties together to form and manage collaborations. For example, place-based engagement helps neighbors form new community groups or block clubs, clean up dirty streets and parks, and track areas of mutual concern such as home burglaries.

Place matters in cities. Civic participation strengthens connections between neighbors, creating a virtuous cycle in which participation creates optimism and coordinated action in the neighborhood, which in turn creates more participation. We can measure this progress not just by the thickening of civic infrastructure but also by the concrete results produced. The path to an effective collaboration doesn't begin with a map, an algorithm, or some elegant data-smart policy solution; it begins with people—residents, public officials, and nonprofit leaders endeavoring to solve a problem. Individuals can marshal support when they tell stories about their needs or unique situations and tell them in a way that draws others in.

Community activists, city planners, residents, and nonprofit leaders often attend town meetings with their own perception of the facts through the lens of their situation or professional discipline. Prioritizing tasks is nearly impossible unless people can agree on what constitutes reality on the ground. For example, a traffic manual may suggest how to improve traffic flow, when neighbors just want a quieter block. Or a declining student population may lead public officials to close a school rather than examine why families are moving out of the community or sending their children to schools farther away. Consider the visual difference between a spreadsheet of the number of opioid overdoses compared to a mapped block-by-block depiction of 911 calls, concentrations of prescription opioids dispensed, capacity of treatment facilities, and so on. Only a compelling visualization of reality will change people's minds.

Maps show linkages, trends, and comparison to other areas, all of which can challenge conventionally held points of view, creating a shared (and often negotiated) understanding of the symptoms, causes, and potential solution sets for a given problem.

Maps with different layers of information serve as platforms for further collaboration. For example, a map showing the sanitary conditions of streets and sidewalks block by block and the city's comparative response by neighborhood, as shown in figure 2.1, in Los Angeles will drive place-based corrective action when the facts are presented for all to see.

Formation: The role of social capital and place

Civic engagement strengthens social capital because it creates greater economic opportunity, neighborhood stability, happiness, and deepened reciprocal trust. For example, neighbors might unite to push their city to set aside a public park where their children can play, or share information on safe walking routes to school. University of Wisconsin professor Chris Holtkamp explores "the relationship of place and identity as an indicator of social capital in its own right." He adds, "Social capital can be defined as the networks and relationships among members of a community expressed through norms of behavior including altruism, trust, and reciprocity."[1] Increasing social capital will produce deeper, more effective, and more numerous cross-sector collaborations. An article by Sohrab Rahimi and others in A Geographic Information System (GIS)-Based Analysis of Social Capital Data: Landscape Factors That Correlate with Trust supports the proposition that "geographic context has a significant association with overall trust."[2]

Figure 2.1. This map, the Clean Streets Index Grid 2016, highlights problem streets throughout the city of Los Angeles. Green-colored parcels help users see which streets are in the best condition, while parcels that scale up from yellow, orange, and red help draw attention to the parcels with the worst street and sidewalk conditions that need targeted intervention.

In his book *Great American City*, Harvard professor Robert Sampson says that when neighbors talk with each other, life runs more smoothly. Leaders of cross-sector collaborations could follow seven of Sampson's 10 principles of social inquiry to produce better community outcomes:

1. Relentlessly focus on context, especially neighborhood inequality and social differentiation.
2. Study neighborhood-level variations and adopt a systematic method of data collection that relies on multiple methods with public standards of measurement.
3. Focus on social-interactional, social-psychological, organizational, and cultural mechanisms of city life rather than just individual attributes or traditional compositional features such as racial makeup and class.
4. Study dynamic processes of neighborhood structural change.
5. Simultaneously assess mechanisms of neighborhood social reproduction and cultural continuity.
6. [Go] beyond both the individual and the local to examine spatial mechanisms that cross neighborhood boundaries.
7. Never lose sight of human concerns with public affairs and the improvement of city life—develop implications for community-level interventions.[3]

Place-based social capital increases the ability of the community to act collectively. And communities with more social capital benefit from these positive outcomes through greater economic advancement, lower crime rates, higher rates of education, and healthier populations.[4] Conversely, low social capital in challenged neighborhoods can lead to adverse results, including more child abuse, less economic mobility, and less happiness than in similar communities.[5]

These compelling findings present an important role for local government in promoting cohesive communities and optimistic residents. According to experts at Pennsylvania State University, characteristics of sustainable neighborhoods include investment in the local community and care of its residential properties. Developed open spaces, such as yards and parks, also may benefit from social capital formation. Finally, amenities such as libraries, museums, and schools may help individuals develop social contacts within their community and increase their overall levels of trust.[6]

The American Planning Association (APA) looked at how to build social capital through urban design and planning.[7] The APA suggests including these design elements:

- Places that provide opportunity for networking and participation
- Traffic management and urban design to create safe places for children to play
- Encourage mingling through place making
- School and classroom size
- Open spaces, (pocket) parks, and yards
- Street condition
- The reduction of one-way streets and traffic speed
- Grass, trees, flowers, and convenient places for outside sitting
- "Eyes on the street" buildings with street-level windows
- Home ownership
- Absence of vacant homes
- Shorter commute times or the opportunity to walk or bike to work
- Mixed-use infrastructure

The APA emphasizes the connection between urban design and social interaction to demonstrate how design, shape, form, and qualities of streets and urban spaces affect how people use them.

If social capital produces better health, more happiness, and economic mobility, then clearly, city officials should focus land use and street and building design efforts on building social capital. One example is the Front Porch Alliance in Indianapolis. The alliance mapped assets such as nonprofits, local businesses, and faith-based organizations in stressed neighborhoods and then structured collaborations to increase social capital in those targeted communities. It involved planning where new sidewalks would increase social interaction or make a child's walk to school safer, and where donated equipment and volunteers would turn a vacant lot where people used crack cocaine into a small playground. The Front

Porch Alliance considered how the city could support community leadership and enhance its credibility by responding quickly to place-based services requests. Similar activities occur in many American communities today.

Using civic engagement to inspire collaboration

In addition to creating more social capital, collaborations that include civic engagement also can inspire action around a specific goal or campaign. To do this, leaders do not start with a blank slate; rather, their organizing efforts take place in the context of these existing conditions:

- Demographic factors such as race and poverty
- The normative and regulatory environment in which the potential participating organizations exist
- Previous failures of a single sector or organization to ameliorate the problem on its own
- Existing relationships among the potential players, which influence trust and the willingness of potential actors to participate with each other
- The existence of an accepted brokering organization or legitimate convener

An organized collective response must be well visualized, easily understood, and iterative. Whether groups form into collective agency is "not merely an issue of structural arrangements or antecedent conditions; it is a process of emergence resulting from communication processes that are distinct from market or hierarchical mechanisms of control."

Elected leaders need the capacity to call the public together to solve an important civic matter. Showing a map can provide a compelling tool for rallying support, whether it's a mayor advancing city aspirations or a community member leading a call to action.

In some cases, stories made in ArcGIS StoryMaps have depicted problems in a way that brings people together to solve complex problems. For example, the San Francisco map of wealth divides helped build the case for affordable housing by highlighting the needs of communities that do not share the prosperity otherwise found in the Bay Area.

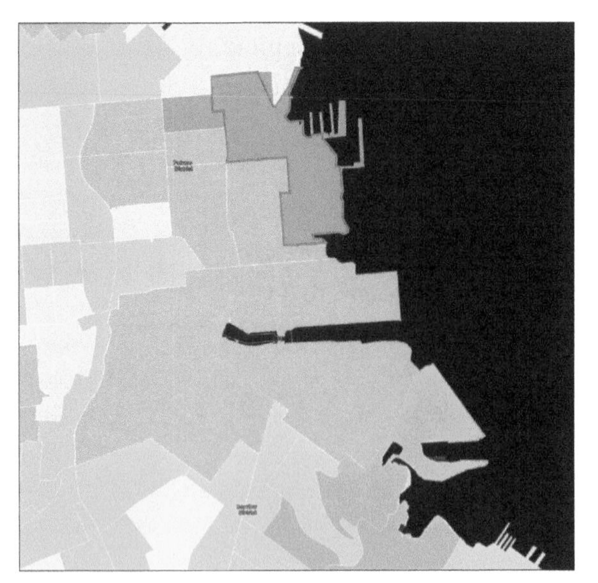

Figures 2.2 and 2.3. The San Francisco map of wealth divides highlights disparities in household incomes in the Bay Area. Dark-blue shaded areas indicate higher levels of household income while yellow-shaded areas indicate lower levels of household income. In figure 2.2, *left*, the blue area marked with a red border is the former working-class neighborhood of Potrero Hill with a median household income of $179,806. In figure 2.3, *right*, the yellow area marked with a red border identifies a nearby neighborhood with a median household income of $16,703.

In another example, the story *Celebrating the Lost Loved Ones* illustrates the enormous personal cost of opioid addictions for families in a community and plays a part in generating coordinated responses to these tragedies. Cincinnati, Ohio, and the surrounding towns in Hamilton County mapped opioid overdose hot spots and created a publicly accessible dashboard showing near real-time updates of opioid-overdose-related emergency services calls. These real-time maps help first responders and citizens groups respond more quickly and strategically. The release of the dashboard has resulted in a 31 percent decline in EMS responses in Cincinnati, while Hamilton County reported a 42 percent reduction in emergency room visits.[8]

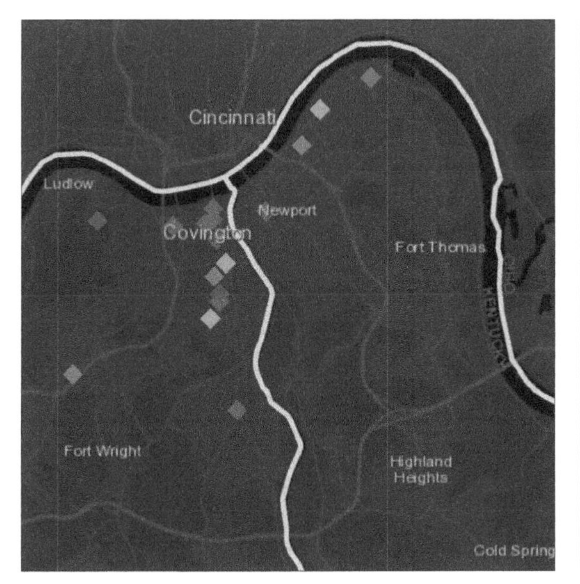

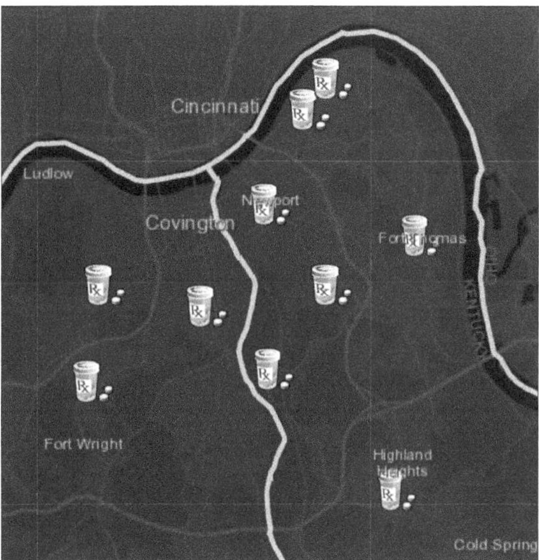

Figures 2.4 and 2.5. These maps highlight locations of opioid treatment facilities and drug drop-off spots (*left*) and syringe access exchange locations (*right*) in northern Kentucky. These maps help area residents find locations where they can access resources and get the help they need.

Defining the problem

Effective action requires a shared understanding and a grasp of the seriousness of the problem. For example, urban growth and development often disguise dense pockets of poverty, and people whose work doesn't take them to impoverished neighborhoods may not understand the density, extent, or inequity of concentrated hopelessness.

Maps are a good way to create shared understanding. For example, when *CityLab* wanted to direct action to the educational achievement gap, it posted an article featuring colorful dot maps that visualized the educational achievement gap experienced by immigrants in Chicago and New York City. *CityLab* associated those findings with racial and economic segregation by neighborhood. The maps revealed the stark differences in educational levels across urban and rural areas and the effects of racial segregation within cities. Cities expert Richard Florida presented rural variations by referencing similar maps.[9]

Harvard University economist Raj Chetty has focused national attention on the consequences of place. Chetty's Opportunity Index demonstrates that the neighborhood in which people grow up affects their future earnings, longevity,

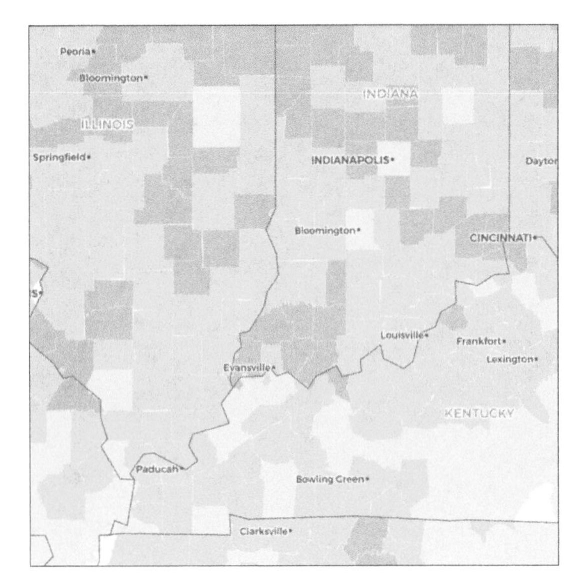

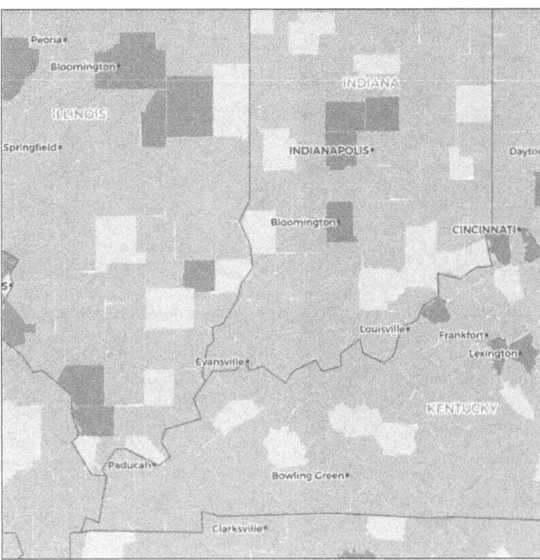

Figures 2.6 and 2.7. Taken from the Opportunity Index, levels of education are shown by parcel on a green scale, *left*, with darker green indicating higher levels of education. Levels of household income are shown on a purple scale, *right*, with darker purple indicating higher average household incomes. These maps help people draw links between education and economic outcomes.

and educational outcomes—all of which are interconnected. The quality and compelling story presented by his maps has not only magnified the importance of his findings but induced action from elected, corporate, and philanthropic leaders.[10]

Even when various parties broadly share interests and live in the same neighborhood, they do not necessarily share a common understanding of the problem. For example, homelessness may be an issue of eviction to one group, domestic violence to another, and addiction to a third. Layered, mapped data helps merge disparate points of view that may exist by creating a shared context for analysis and understanding. This process may involve various visualized hypotheses on the path to defining a common problem, which increases the likelihood of agreement to further develop solutions.

For instance, Louisville's campaign to improve credit access for *Redlining Louisville* Black residents began in 2011 with considering the impact of historical financial obstacles, which prompted a shared understanding about the role of discrimination in finance. Los Angeles engaged with community groups to combat evictions with the aid of its *Anti-Eviction Mapping Project,*[11] which produced collective action against illegal landlord actions. The maps produced in these and other cases related to equity can drive action in the formation phase or they can be part of the foundation of collective cross-sector action.

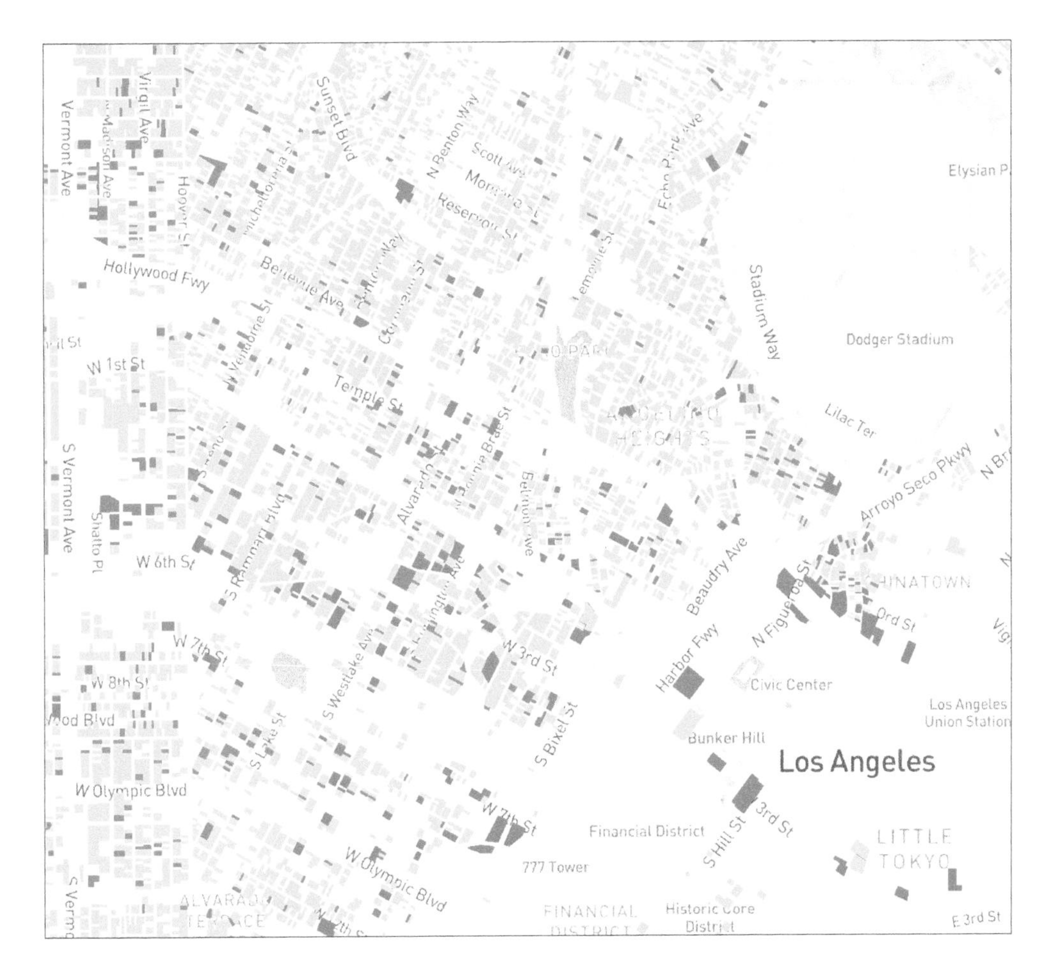

Figure 2.8. The Anti-Eviction Mapping Project in Los Angeles built this map to help voters see the impact of repealing the state Costa Hawkins Rental Housing Act on multifamily units. Blue areas show areas covered by rent control laws, and bright purple areas highlight areas that would be eligible for rent control if the Costa Hawkins Rental Housing Act was repealed.

Geographical analysis exposes challenges compounded by issues of race and place. In *Great American City,* Sampson reinforces the significance of racial segregation, "as the spatial isolation of African Americans produces exposure to multiple strands of resource deprivation, especially poverty and single-parent families with children. Ecologically concentrated disadvantage is a powerful form of disparity predicting both individual outcomes and variations in rates of behavior across neighborhoods."[12]

Issues of structural poverty require effective systemic responses instead of responses by a single agency or organization. Mapping a specific area facilitates the identification of partners, the arrangement of their interventions, and the collection of location intelligence.

One can see the linkage of planning, mapping, and community action in the campaign to improve nutrition in Paterson, New Jersey. A Nutritional Environmental Measures Survey of the city's First Ward corner stores documented deficiencies in nutritional offerings. The United Way, in its role as organizer of a collaborative response, needed to answer the question: What groups should participate to improve fresh food offerings?

The United Way used mapping to identify assets supporting food access within the city, while also examining the relationship between transit and food access. This analysis led to a broad-based coalition of almost 20 nonprofit organizations and public agencies that eventually formed a collaboration to address food safety, neighborhood revitalization, and education. The United Way, with support from a local nonprofit, City Together, augmented existing maps with data generated by community groups. These mapping efforts supported the development of specific strategies to improve food access.

The Passaic County Food Policy Council joined with the United Way to identify strategies to support access to healthy foods. Community-based groups that serve minority populations received small grants to help organizations add information to the map about the built environment, safety, and transportation options (for example, bicycling, walking, and safe routes to transit, schools, and parks).[13] Displayed in a GIS format, this information helped the collaborators focus on the right issues in the right places.

In all these examples, the visualizations clarified the problem and provided a foundation on which the collaboration was formed.

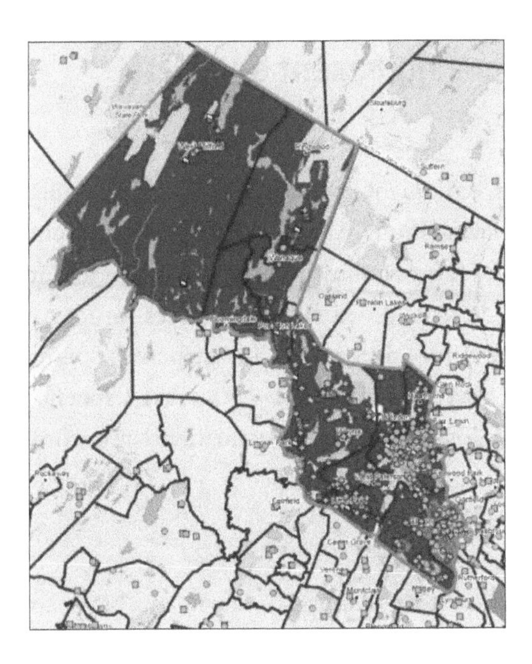

Figure 2.9. This map shows food accessibility for Passaic County, New Jersey, residents in 2009. Green circles indicate limited-service stores, orange squares indicate full-service supermarkets, and blue dots indicate community-supported agriculture.

Interagency collaboration

Mapping can serve as virtual scaffolding between the public and governmental agencies. In Los Angeles, for example, the geocoding of public data enabled through the city's GeoHub facilitated interagency collaboration on reducing homelessness. In Chicago, epidemiologists improve their predictions about the spread of disease by easily accessing mapped data from other agencies.

If partnerships don't develop, opportunities are lost. Take the example of a volunteer working for a national nonprofit and appointed by a court to assist a child under its care. The volunteer needs to connect the child to available services, but since governmental agencies do not share their information geospatially, let alone formally collaborate, they cannot easily find accessible services near the child's home or school or information about the availability, quality, or expertise of those services. Mapping resources with information about expertise and availability increases access and reduces the difficulty of case workers, especially volunteers, in finding the most effective services.

Operations: Five ways civic engagement supports action

Civic engagement may be the end goal, or it might be the means of helping guide the operation in accomplishing a specific mission. Examples of civic engagement at work range from how a city charts its future to how it makes decisions about infrastructure and enforces local ordinances. The next five examples—increasing volunteerism, envisioning a city's future, informing infrastructure spending, enforcing policies, and changing government policy—showcase a specific cross-sector operational model and describe how mapping furthers the accomplishment of its goals.

Volunteerism

Volunteerism constitutes a critical part of the American fabric and plays an important role in strengthening neighborhoods. People often volunteer through organizations or their work, often in conjunction with their local government. Using geospatial data can make it easier to volunteer in a specific place for a specific organization and increase the amount of service accomplished.

One successful example can be found in the work of Bloomberg Philanthropies, through its Cities of Service Love Your Block program. It was designed to help match volunteers with appropriate nonprofits in cities. In Love Your Block, city officials and private and nonprofit workers join to tackle blighted properties, clean up neglected areas, and create community gathering spaces—block by block. From 2009 to 2018, "more than 10,000 volunteers have been engaged to remove over 480,000 pounds of trash, clean up nearly 600 lots, and create more than 180 art displays, in addition to numerous other community projects" in 10 cities across the US.[14]

Additionally, the Urban Institute found that Love Your Block also produced more follow-on engagement and stronger civic ties.[15] An operating goal of Cities of Service is to provide the infrastructure to make it simpler to find the right place to volunteer. Mapping volunteer opportunities helps identify these choices. This collaboration can improve volunteer activities, along with nonprofit and government resources, to address important urban problems.

Envisioning the future of a city

Some cities form collaborations to help chart their future. In 2017, the City of Brampton, Ontario, Canada, began an effort to think more broadly about its future and envision what the city could be like in 5, 10, 25 years, and beyond. This effort engaged multiple stakeholders in the development of a long-range plan, with community engagement and the vision as the desired outcome. Brampton used geospatial visualization as one of several tools to achieve its aims. Residents used a mapping tool to highlight what they liked about the area ("I love it here"), what they wished they had in the area ("I wish this was here"), or what they thought the city could do better ("We can do better here").[16] Additional efforts included meetings and workshops to increase engagement.

Mapping Brampton's future allowed residents of all ages, backgrounds, and experiences to share their ideas. Brampton collected input from these areas and crafted a vision for the city:

- Age-friendly strategy
- Cultural heritage master plan
- Economic development master plan
- Environmental master plan
- Official plan
- Plan for affordable and accessible housing options
- Protect employment lands
- Queen Street rapid transit corridor master plan
- Sustainable development guidelines and thresholds
- Trails and pathways, active transportation and cycling strategy
- Transportation master plan

As stated by Matt Leger, writing in *Data-Smart City Solutions*:

> Mapping enabled city residents to participate in greater numbers than ever before by providing a platform for in-person and online idea sharing. Brampton's residents were truly engaged when they could hover over a map as a team and visualize their ideas for the future. During the roundtable

> process, mapping enabled critical stakeholders to collaborate and helped identify areas where government, business, and community resources could be leveraged most effectively to implement the [action plan]. Lastly, mapping aided city planners in prioritizing which areas of the city were in greatest need of investment, helping bring resources first to citizens who need it most.[17]

Winnipeg used a similar approach to produce a vision for its downtown. The city presented maps and asked for individuals, organizations, and businesses to offer their ideas on safety and livability. Winnipeg's map, *Urban Brew: Planning a Safer Downtown,* is the operating structure for participatory feedback. The map furthers multiparty cooperation and a social exchange of ideas that benefits from interactivity, produces more trust in the outcome, and serves as a foundation when the effort moves from vision to implementation.

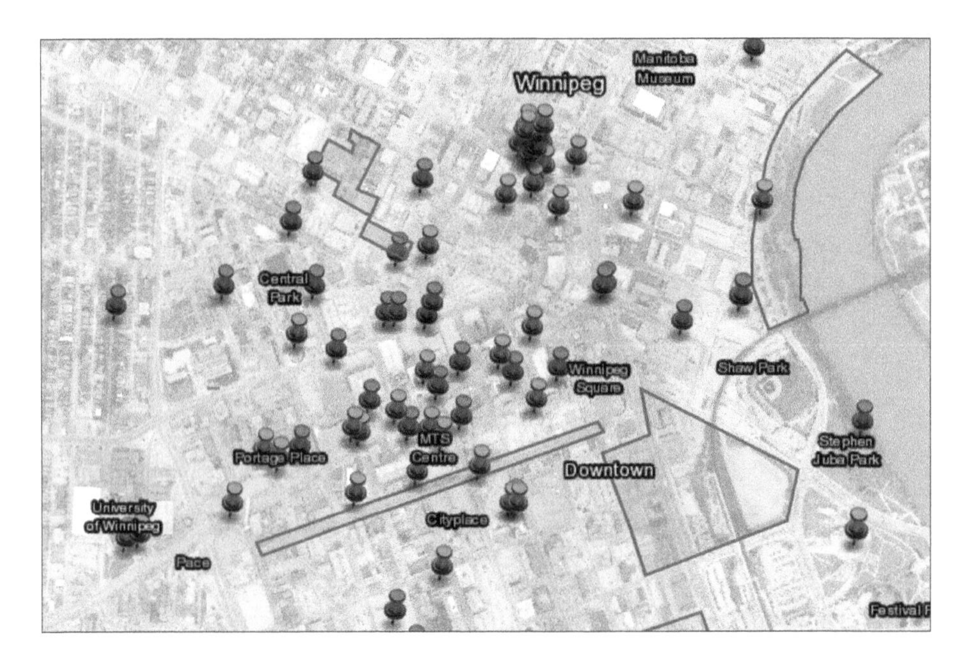

Figure 2.10. In this map, green pins indicate where Winnipeg residents feel safe. Red pins indicate areas where residents feel unsafe. This map was part of a larger discussion that also asked participants where they live, whether they work downtown, and how they think safe downtown spaces should be designed.

Informing infrastructure spending

Infrastructure investments have serious consequences for communities, bringing with them substantial benefits and short-term costs. When planning where a construction project occurs and how it is configured and executed, the community benefits when more stakeholders participate in the discussion. Yet broad participation depends on the quality of the GIS visualization platform and its ability to present meaningful options.

The Regional Transportation Alliance of South West Pennsylvania approached planning for the future of better transportation by aiming to collaborate as broadly as possible. To gather ideas, the alliance provided residents with a set of interactive maps of the city's transportation network and asked them to help answer the following questions: What do we want from our transportation system in the next 10 to 20 years, and how do we get there? Their online report, *Imagine Transportation 2.0, A Vision for a Better Transportation Future,* uses interactive maps, links to case studies, and references to thought leaders and futurists to engage hundreds of community groups, businesses, and local leaders across the 10-county region.

In Fort Lauderdale, Florida, a collaborative group came together to assist community members, homeowners, and businesses in determining what individual actions they could take to further the city's sustainability efforts. The *Green Your Routine* map features stories and identifies specific places where "City leadership, staff, neighbors, businesses, schools, and community organizations are addressing climate challenges; conserving energy and resources; building to new standards; protecting and preserving our air, water, and natural environment; considering transportation choices; and recycling and reducing waste."

The Fort Lauderdale GIS staff provided updates highlighting user actions, including maps showing possible composting sites, electric car-charging stations, and ways to implement renewable energy options. Unfortunately, the Green Your Routine initiative demonstrated the important challenge of updating maps so they can remain useful. In Fort Lauderdale, staff had difficulty obtaining funding to maintain the map project, thus reducing its overall effect—an issue discussed further in chapter 7.

Civic engagement organized around maps can drive better investment decisions for infrastructure. City officials as well as neighborhood and business groups use mapping to identify priority areas for investment. The City of Kitchener in Ontario, Canada, presented neighborhood-specific maps and pictures to support

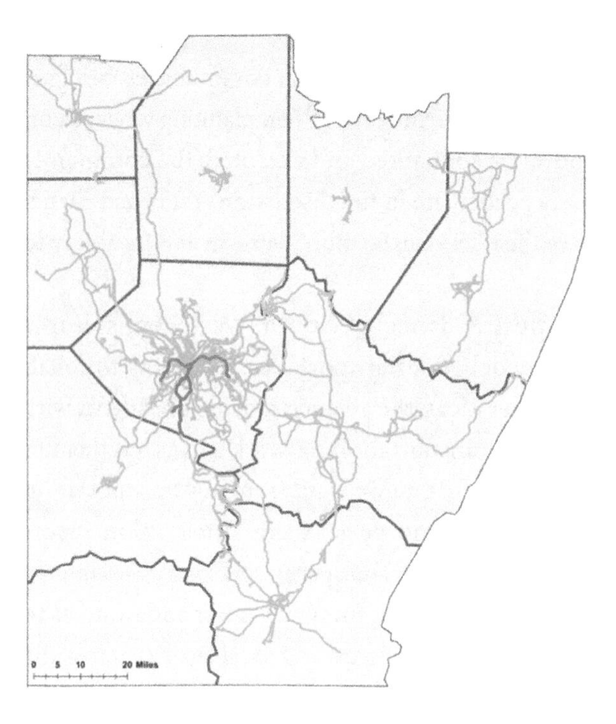

Figure 2.11. This map from the Southwestern Pennsylvania Commission shows the concentration of investments on transit routes in Allegheny County in 2015, where the total operating funds expended on fixed routes was $330,795,423 (routes shown in purple) out of $358,118,989 spent on the area shown.

investments. Using these visuals, organized by neighborhood, residents could picture what those investments might provide and offer feedback.

As mayor of South Bend, Indiana, Pete Buttigieg oversaw a $60 million overhaul of the city's parks. Before he spent the money, Buttigieg faced questions of how and where to use the new resources. In response, he created a process with visualized proposals so that residents could connect with community-based organizations, the city, and each other by using a map-based platform. With the aid of The Trust for Public Land, the mayor and his staff posed a series of questions about the ease of access to city parks and the condition of each property.

More specifically, these questions asked about parking and walkability and relationships to neighborhood wealth. Maps helped groups visualize areas that needed improvements and other basics concerning the built environment. The city GIS team allowed residents to augment the maps with pictures and comments about what they wanted in the parks. Posted signs in the parks invited community comments through either a phone call or a data response to the city through its My South Bend Parks & Trails initiative.

These responses provided a substantial amount of spatial intelligence, including insights about park users' ages and whether park visitors thought equity and design were adequately addressed. In this effort, civic engagement was both an input and a desired outcome of the collaboration designed to inform city investments.[18]

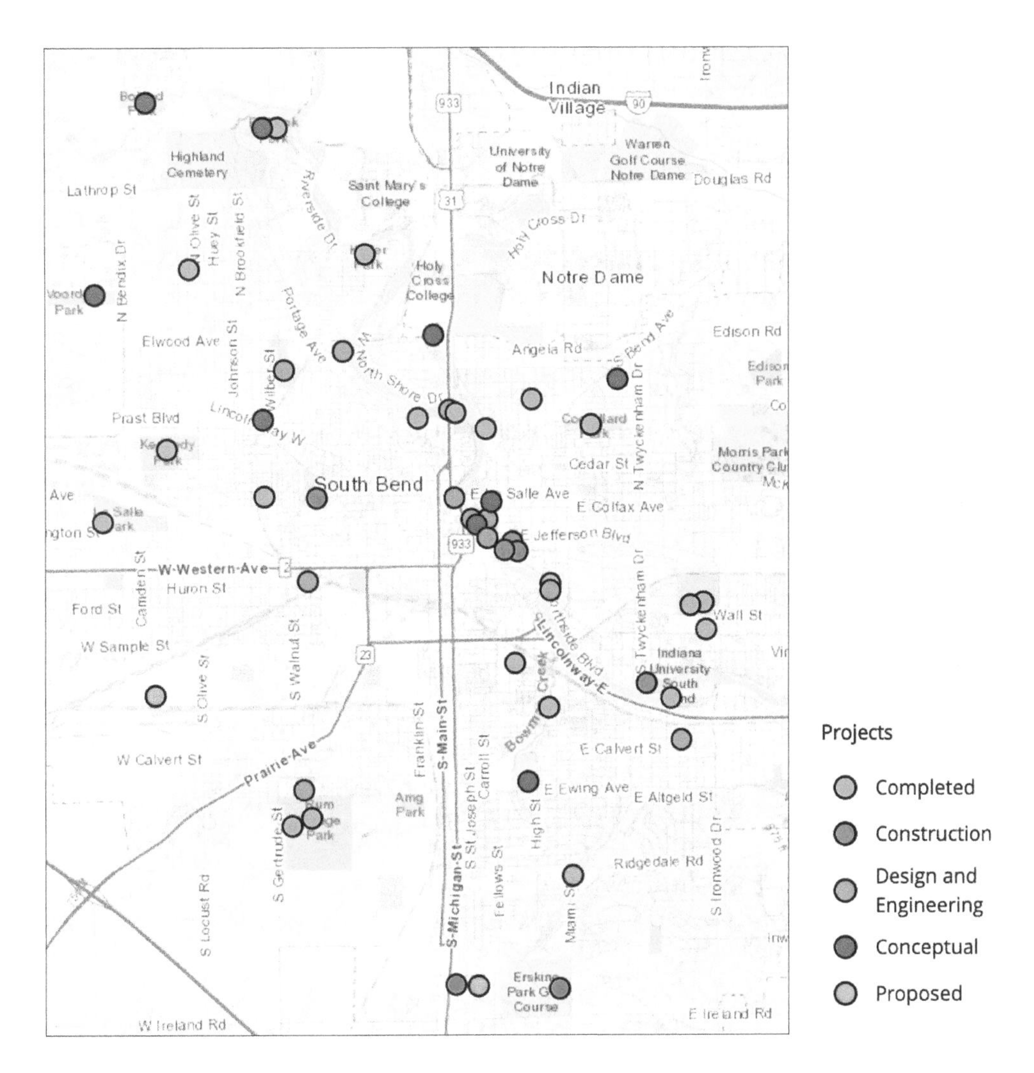

Figure 2.12. This map shows all past, present, and future projects that are part of the My South Bend Parks & Trails initiative. Map users can click each dot and get more details about that specific project.

Major city infrastructure projects promise long-term benefit but often cause residents short-term aggravation. For example, the New York City Second Avenue Subway construction project created noise and congestion, obstructed entrances to apartments, and hurt retail business. What would have happened if the city had allowed feedback on the plans in an interactive map? Washington, DC, provides an instructive example.

In March 2016, Washington Metropolitan Area Transit Authority (WMATA) announced that it might close the Metro's Blue Line for six months to make critical repairs. In response, Esri produced an interactive story that helped people visualize the location and impact of the closures and offered individuals, businesses, and community groups the chance to collaborate on mitigation strategies.

The Esri DC Blue Line story was designed to facilitate a discussion about the impact on the 18,000 businesses in the affected area and the $60 billion in annual revenue they generate. The process involved residents, business, and government officials concerned about various issues, from how their business would survive to the loss of tax revenue to practical ways to soften the effect. Posing specific

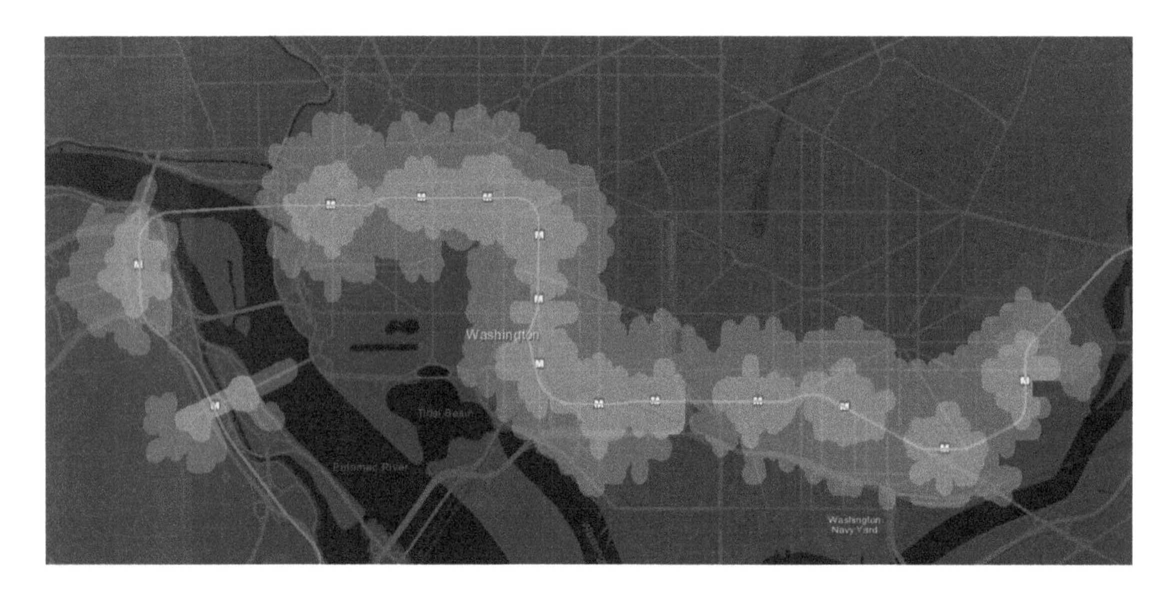

Figure 2.13. The shaded areas on this map represent businesses and people within a quarter-mile walking (pink) and half-mile driving (red) distance of Washington, DC's Metro Blue Line. The map helps people understand how Metro closures can dramatically impact commute times, business revenue, employment, rents, and property value.

questions about the effect on home values, landlords, or transit options allowed people to suggest solutions such as new bus routes. The story helped citizens envision what might occur and facilitated a multiparty discussion of ways to address various issues.

In these examples, cross-sector collaborations, powered by maps, support the process of determining where and how cities should use their infrastructure dollars, modifications to design, and efforts to alleviate localized disruption.

Aiding enforcement of laws

City officials can organize collaborative civic engagement to extend the reach of important city services, something academics sometimes call the "coproduction of public goods." These organized partnerships increase when residents use their smartphones to take location-stamped photographs and share the images and related information with others on a social media or city site. Such mechanisms in San Francisco and Chicago allow city food inspectors to analyze location-based comments more quickly from diners to identify health and sanitary conditions and food-borne illnesses in area restaurants. By encouraging civic engagement that involves geotagged input from residents, the inspectors can increase their speed and extend their reach.

New Orleans and Detroit also applied this platform approach of mapped open data. Both faced serious problems of blight and limited resources to address it. They encouraged residents to submit pictures of vacant or blighted homes. The cities uploaded these pictures onto maps to help official inspectors identify blight early on. This approach helped preempt more serious problems with blighted and adjacent structures. Engaging residents in blighted housing enforcement helped New Orleans and Detroit officials augment their public resources and provided an entry point to civic action.

In the United Kingdom, the term *fly-tipping* describes illegal dumping of construction rubble, tires, mattresses, and so on. Because enforcers cannot stay ahead of the problem, officials encourage citizens to add their observations to target-area maps. The United Kingdom's *What A Tip!* map shows specific problem areas combined with map layers showing disposal sites, population data, and other relevant information. The map supports civic engagement through a platform that allows people to report incidents easily to local authorities. The approach is an interesting linkage of technologies: well-visualized layered maps tell a story and lead users to Fix My Street, a website for uploading photographs

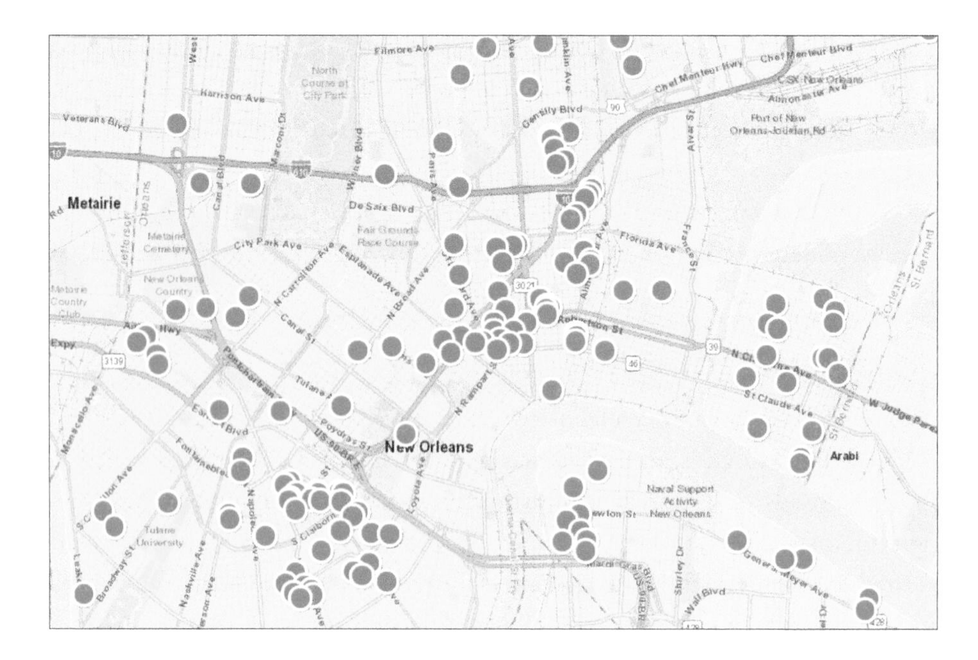

Figure 2.14. New Orleans officials built the interactive BlightSTATUS map to support engagement efforts. The map's tools allow users to filter by time frame of public nuisances and blight in the city. In this snapshot, each red dot represents an instance of reported blight. The map shows less than a quarter of the instances of reported blight across the city during a three-month period.

and reporting the postcodes, locations, and details of problems related to any street in the United Kingdom.

Changing government policy

The previous examples highlighted situations in which governments partnered with other sectors and residents to bolster an existing public service. In other instances, however, collaborations form with the goal of encouraging governments to address service needs. They form outside of the government and often use mapping to advocate for and force public action. For example, the Dallas Children at Risk and the Food Trust project was designed to pressure the city to address food insecurity for children. Maps created for the project showed that 36 percent of Dallas County zip codes contained areas with limited access to affordable and nutritious food. Other maps showed travel distances to the locations of fresh food, supermarkets, and variety stores. Those factors mapped against obesity rates showed the need for more equitable results.[19] The study concluded that

Dallas is well positioned to adopt a program such as the New Orleans Fresh Food Retailer Initiative or the Philadelphia Healthy Corner Store Initiative to increase access to healthier food choices.

The impact of the initiative lowered the level of food insecurity in Dallas between 2013 and 2017.

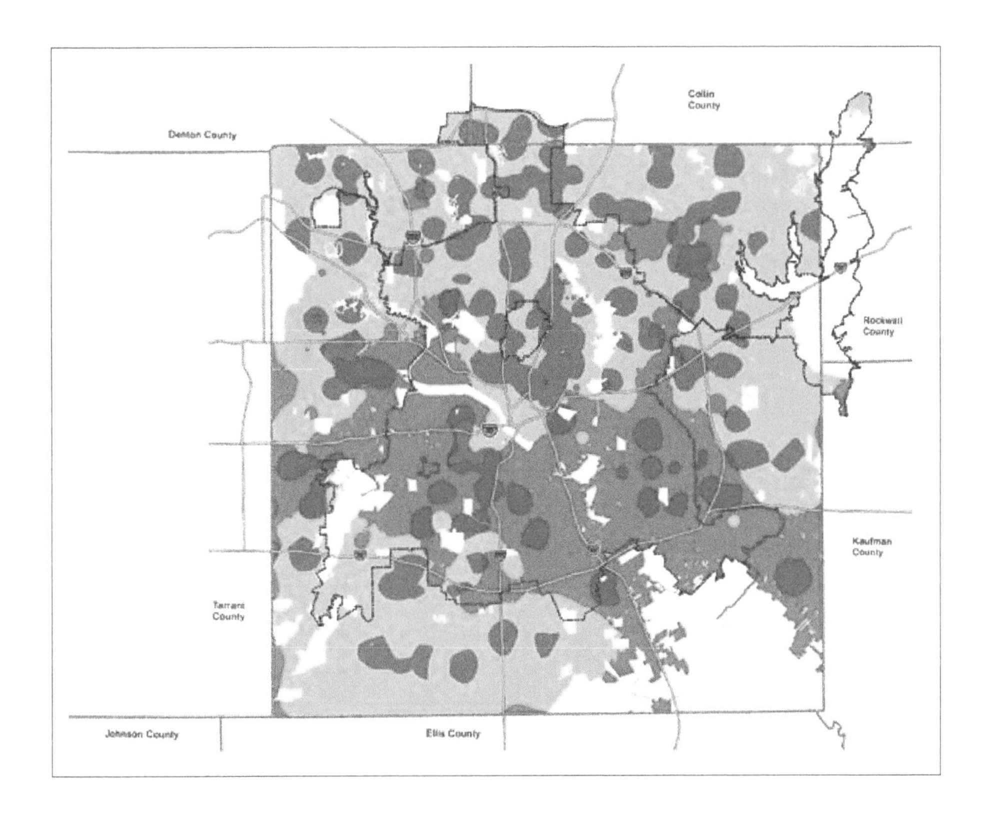

Figure 2.15. Supermarket sales and income show where Dallas County supermarkets do not adequately serve lower-income communities (highlighted in red) and areas that have fewer supermarkets, regardless of income levels (yellow).

Adaptation

Civic activism can drive the formation of a collaboration and become part of the operation. Once underway, successful collaborations use acquired knowledge and new tools to adapt their approaches as they learn what works. One example of adaptation is the work of the Dolores Huerta Foundation, founded by namesake Dolores Huerta, a civil rights activist, advocate for immigrant and worker rights, and cofounder with Cesar Chavez of the United Farm Workers union. Her foundation focuses on civic engagement and education. Its neighborhood-based work involves several chapters and hundreds of community members. According to Camilla Chavez, executive director of the foundation, the foundation has helped secure millions of dollars in sidewalk, streetlight, and sewer investments. It used map-based platforms in its *Vecinos Unidos*, or Neighbors United, initiative to guide and coach community members on how to navigate political systems.

The foundation started with hand-drawn maps before it had access to digital tools. In recent years, GIS analyst Sophia Garcia has converted the foundation's hand-drawn maps to high-quality digital versions that highlight infrastructure gaps and inequities in economic conditions and educational outcomes. Garcia engaged college students with GIS skills to show neighborhoods where particularly poor road conditions endangered pedestrians and bicyclists and used the information to advocate for needed street improvements. These adaptations produced even more effective place-based advocacy as Chavez tasked volunteers and staff to ride streets in the targeted area, take pictures, and upload them to maps. As Garcia explains, "It is really hard for any elected officials to refute the data and the maps that we put in front of them."

The foundation's expanded capacity for GIS mapping and its successful tactics allowed the organization and its partners to adapt and evolve into other areas, such as the census Complete Count Committees and the South Kern Building Healthy Communities initiative. Chavez explains that they use maps when meeting with community members to clarify a new problem and organize the groups into action. Chavez noted one example, when the foundation challenged a school district on the issue of gerrymandering. "We used the maps to compare where the school board trustees lived to where the Latino population lived, which rallied people a lot," she said. The foundation's Mapping Social Justice in Kern County narrative asked the question: "Is this really representation?" The school district eventually redrew the gerrymandered school district boundaries. A combination of success, new collaborative partners, and new tactics allowed the foundation to constantly adapt and take on additional important issues.

Maps produce the common bridge and understanding across languages for the foundation. Disadvantaged groups may not be heard if they try to communicate these concerns to their officials without visual data to back them up, foundation officials explained. The foundation showed that map-based engagement allows a collaborative to adapt its fundamental goals and operations to tackle new problems, often applying the techniques and tools in an earlier stage to accomplish new missions.

Lessons for collaboration: Turning engagement into results

Government and nonprofit organizations build social and political capital when they inform the public about their efforts; transparency, in turn, builds confidence. Ongoing access to information about what is occurring where one lives or works increases trust, and dynamic maps are an effective and transparent means to keep interested parties informed about actions taken. People who see their words translated into action will stay involved, producing social capital and involvement that enhances community health and safety.

A. O. Hirschman in his classic book, *Exit, Voice, and Loyalty*, argues that people invest in their communities, financially and socially, when they believe their voice matters; and when they believe they have no voice in their communities, they exit.[20]

Clearly mapping a call to action, followed by tracked and well-visualized progress, can increase people's confidence that their voice matters. Even when a city or nonprofit collaboration has a large community footprint, most residents remain unaware of results or even the activities designed to produce results that affect them. The continuing flow of information encourages participation. People who participate provide insights and suggestions to the government and nonprofits that bring resources to the community. Civic engagement—voice—builds trust and a stake in the community when one sees and monitors the result. Well-visualized maps constitute the best way to see it.

Notes

1. Christopher Holtkamp, "Social Capital, Place Identity, and Economic Conditions in Appalachia," May 2018, https://digital.library.txstate.edu/bitstream/handle/10877/7290/HOLTKAMP-DISSERTATION-2018.pdf?sequence=1&isAllowed=y.

2. Sohrab Rahimi and others, 2017.

3. Robert J. Sampson, *Great American City: Chicago and the Enduring Neighborhood Effect*, 2012: Kindle version Location: 6,117, https://www.press.uchicago.edu/ucp/books/book/chicago/G/bo5514383.html.

4. Washington APA's Game Changing Initiative Social Capital Working Group, "Building Social Capital Through Urban Design and Planning Activities," 5, https://www.washington-apa.org/assets/docs/2015/Ten_Big_Ideas/21_%20updated%20use%20this%20version%20big%20ideas%20social%20capital%20report.pdf.

5. John M. Bryson, Barbara C. Crosby, and Melissa Middleton Stone, University of Minnesota, "The Design and Implementation of Cross-Sector Collaborations: Propositions from the Literature," https://www.jstor.org/stable/4096569.

6. Matthew A. Koschmann, Timothy R. Kuhn, and Michael D. Pfarrer, "A Communicative Framework of Value in Cross-sector Partnerships," *The Academy of Management Review* 37, no. 3 (July 2012): 332–354, https://www.jstor.org/stable/23218092.

7. Jane Wiseman, "Data-Driven Approaches to Fighting the Opioid Crisis" (unpublished paper, February 2019).

8. Tanvi Misra, "Mapping the Achievement Gap," *CityLab*, March 13, 2017, and interactive visualization created by Kyle Walker, a geography professor at Texas Christian University, https://www.citylab.com/equity/2017/03/a-us-dot-map-of-educational-achievement-segregation/519063/?utm_source=facebook&utm_term=2019-01-17T16%3A33%3A31&utm_campaign=citylab&utm_medium=social&utm_content=edit-promo.

9. Richard Florida, "Some Rural Areas Are Better for Economic Mobility," *CityLab*, October 2, 2018, https://www.citylab.com/equity/2018/10/rural-areas-are-better-economic-mobility/571840.

10. R Chetty, "Opportunity Index," https://opportunityindex.org.

11. The Eviction Defense Collaborative (EDC) and The Anti-Eviction Mapping Project, https://www.arcgis.com/apps/MapJournal/index.html?appid=e9d1638ec7724e899325e88ad62d4089.

12. Sampson, location 6222.

13. Together North Jersey NGO Micro-Grant Project, *Food Environment in the First Ward of Paterson, NJ*, https://www.dropbox.com/s/ks812n350suhscs/TNJ-United-Way-Passaic-County-Final-Report_04202015v2.pdf?dl=0.

14. Cities of Service, "Love Your Block Program," https://citiesofservice.jhu.edu/press-release/cities-service-awards-250000-cities-neighborhood-revitalization.

15. Ibid.

16. Matt Leger, *Data-Smart City Solutions*, https://datasmart.ash.harvard.edu/news/article/map-monday-bramptons-vision-future.

17. Ibid, Leger.

18. Steve Goldsmith, *Data-Smart City Solutions*, https://datasmart.ash.harvard.edu/news/article/leading-locational-intelligence.

19. S. Albert, M. Manon, and R. Waldoks, *Food for Every Child: The Need for Healthy Food Retail in the Greater Dallas Area*, (Philadelphia, PA: The Food Trust, 2015).

20. A. O. Hirschman, *Exit, Voice, and Loyalty*, (Harvard Press, 1970), 30.

3

Extending social services

Philanthropists, federal, state, and local agencies fund a broad spectrum of social programs that serve tens of millions of Americans annually. They offer childcare for parents; provide foster care and Head Start for children; serve the elderly, youth, and military families; extend food and rental assistance; and furnish health and mental health services. While the list of programs is long and varied, the list of program providers is even longer. Human services organizations numbering in the hundreds of thousands have made up the single largest category of public charities reporting to the IRS in recent years.[1]

Of course, quantity of effort is not the same as quality of effort. More than 70 years ago, Franklin Delano Roosevelt observed, "The success or failure of any government in the final analysis must be measured by the well-being of its citizens."[2] Since Roosevelt's time, our understanding of well-being has evolved substantially. Research suggests that it depends on many "complex interactions between individual, family, neighborhood, and ... community."[3] We now recognize that neighborhood or place effects influence people's behavior beyond their individual and family characteristics.[4] To affect the individual, we must also affect the place. Yet in too many communities, scores of social services organizations and municipal agencies work in silos, each with an incomplete window into the resources and dynamics at play in the neighborhoods they serve. Because these organizations operate in parallel rather than in concert, we see duplication and gaps in coverage occurring concurrently within a single community. Without collaboration, organizations not only fail to improve the well-being of citizens; they fail to allocate resources in ways that maximize their impact.

This chapter profiles six collaborations that operate from the premise that using information organized around place allows the organizations to more effectively and efficiently serve their communities in ways that improve access, reduce duplication, and enhance outcomes.

Formation

Certain conditions favor the formation of collaborative enterprises. These conditions include the enactment of federal, state, or local mandates; the availability of dedicated or anticipated funding; and the general consensus that one sector or the other fails more than it succeeds in its efforts. However, even in the presence of favorable conditions, collaboration often requires a compelling leader to convince potential partners that they have a stake in working together. In the first case profiled, from Marion County, Indiana, sector failure led to legislation that served as the impetus for a cross-sector initiative. In the second case, also from Marion County, anticipated funding motivated a faith-based organization to reach out to the community and rethink its go-it-alone strategy.

In both cases, potential partners aided by location intelligence conducted the types of planning and analysis typically associated with the formative stage of collaboration. Geolocating risk factors, existing programs, and barriers to access set the stage for their ongoing work together. In the resource-constrained world of social services, duplication of effort in the face of unmet need makes a powerful case for coordination and collaboration.

Making the case for the co-location of social services

The City-County Council of Indianapolis/Marion County established the Early Intervention Planning Council (EIPC) in 2005. This cross-sector council includes representatives from city and county agencies; the justice, school, and health systems; social service organizations; and individuals from the community. EIPC's founding purpose was to plan early intervention services for children and their families at risk of abuse, neglect, and delinquency. Today, EIPC oversees implementation of the county's Early Intervention and Prevention (EIP) Initiative to reduce involvement in the child welfare and justice systems by coordinating the work of services providers in Marion County.[5]

Shanna Martin, former director of EIP and now director of community initiatives at The Children's Museum, describes the formation of EIPC as a response to what was then seen as the failure of the child welfare and juvenile justice systems. To create a better system, the council drew people from many sectors and built around that broad context. The council needed to know what brought kids and their families into contact with the system. "We needed to challenge ourselves about what we should be doing, what we should be focusing on," Martin said.

EIPC first engaged the Marion County Commission on Youth (MCCOY) to help with strategic planning. Today, it is the council's coordinating agency. MCCOY led the development of two plans for EIPC. Research conducted before and during the first planning process confirmed what council members suspected:

- Early intervention and prevention reduce the risk of problems escalating into crises; fewer crises reduce the likelihood of involvement in the child welfare and juvenile justice systems.
- Numerous organizations throughout Marion County provide early intervention and prevention services, but poor coordination often hampers their efforts.
- Transportation and other barriers prevent families from accessing available services.

Based on this research, the council explored co-location of multiple social service providers in a single center so that children and families could more easily access preventive and early intervention services. MCCOY convened a cross-sector advisory committee and engaged SAVI, a Social Assets and Vulnerabilities Indicators program of the Polis Center at Indiana University–Purdue University Indianapolis (IUPUI), to help select a location for the center. Sharon Kandris, associate director of The Polis Center and director of SAVI, describes its work: "We bring together huge amounts of data from over 60 sources and turn it into meaningful, community-level information. We provide ways for people to visualize it so that they can develop a more comprehensive understanding of the challenges and opportunities in their community."

Given that they could only fund one center, the advisory committee asked, "In which Indianapolis neighborhood would a co-location of services center prevent the greatest number of at-risk children from entering the child welfare and juvenile justice systems?"[6] MCCOY and SAVI began their investigation by creating a Children and Family Needs Index. They found that living in poverty, living with a single-parent family, and living in a neighborhood with a high crime rate were among the factors that put children at risk of involvement with the child welfare or juvenile justice systems.[7] From the list of factors, the group selected 12 indicators related to demographics, education, health, housing, public assistance, public safety, and the economy.[8] They assigned weights to each indicator and layered the information onto Marion County census tracts, allowing the committee to visualize and identify areas of extreme need. As seen in figure 3.1, the higher the index value, the greater the need.

Next, SAVI layered neighborhood assets on the community needs index. The gap analysis included more than 30 types of programs delivered through more than 730 locations. The analysis revealed the gaps, despite the diverse array of community resources.

Using gap analyses such as the one shown in figure 3.2, the committee selected five communities for detailed consideration. SAVI added bus routes to the community data to assess travel time, connectivity, and access to resources. The committee then selected two of the five target communities for further study. The maps now included sidewalks, neighborhood associations, and community council areas of influence. Figures 3.3 and 3.4 show the asset maps for the finalists, the Near West community and the Southeast community.

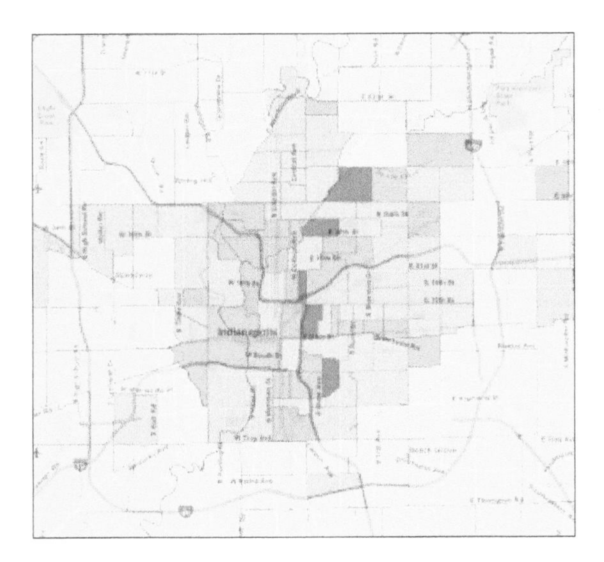

Figure 3.1. SAVI's Children and Family Needs Index shows the areas in extreme need of services—the greater the need, the darker the color.

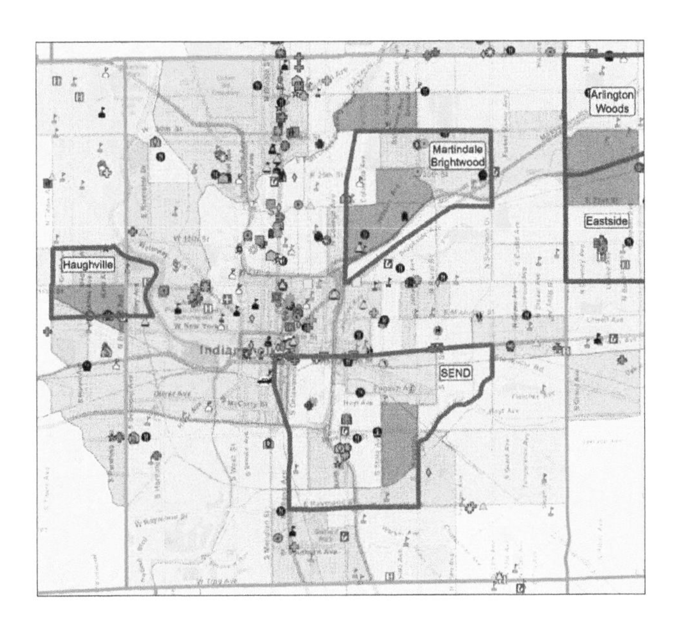

Figure 3.2. Corridors and neighborhoods where Center Township's assets are clustered relative to areas of need. Assets include a wide array of programs such as substance abuse services, schools, family support services, youth and employment programs, and maternal and infant care.

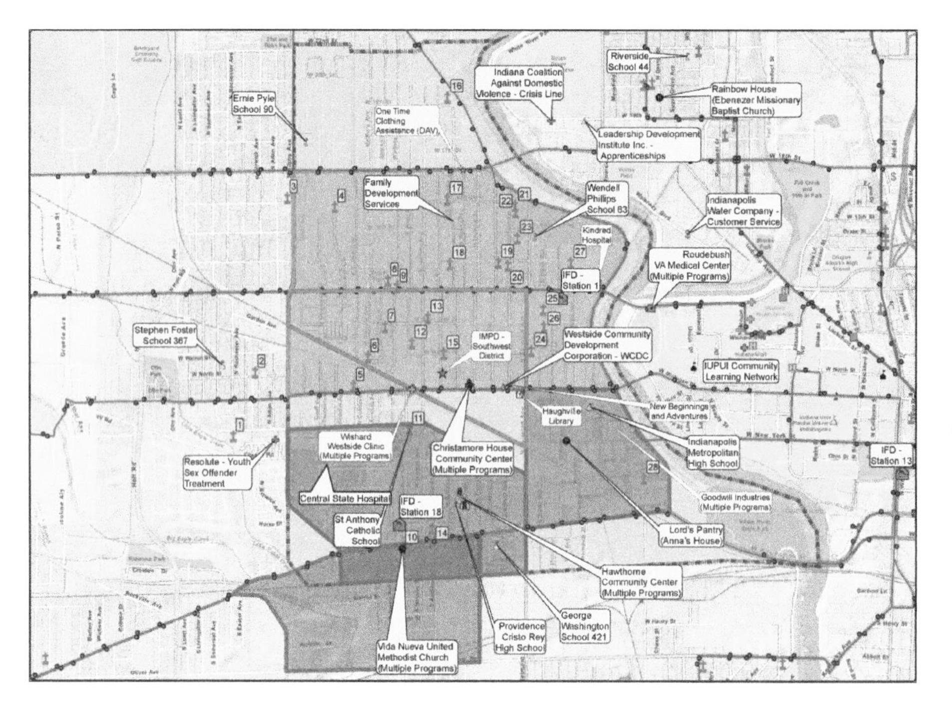

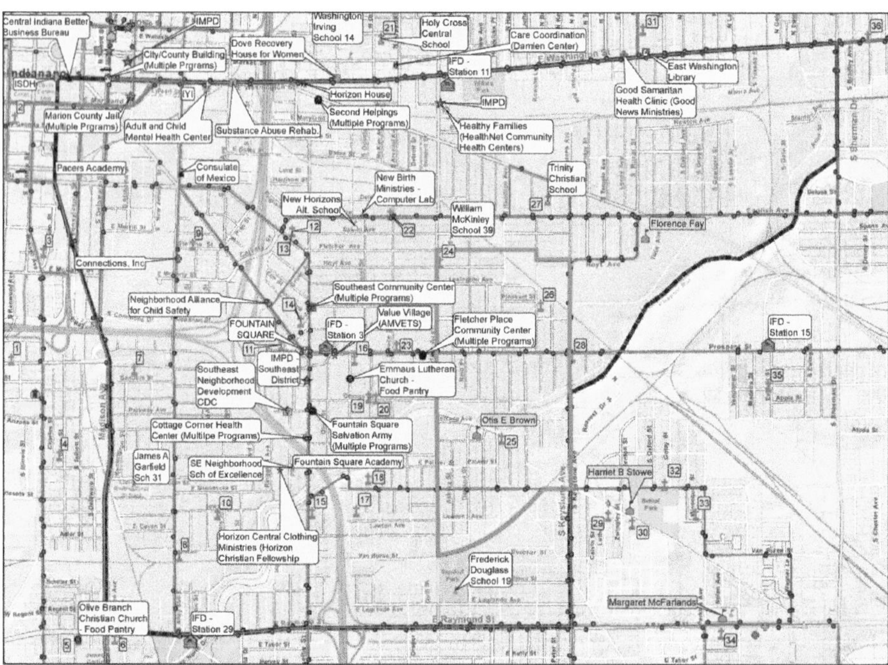

Figure 3.3 shows assets in the Near West community. **Figure 3.4** shows assets in the Southeast community. These maps show how SAVI researchers used detailed asset maps to identify areas of greatest need, outlined in red, and to identify potential partners and sites for co-location of the service center.

The research team also met with community representatives and potential partners and developers to further evaluate the communities. Ultimately, the large and diverse advisory committee, which included government, nonprofit, and community members, selected the Near West community for its planned intervention. Data mapping facilitated the consensus. John Brandon, president of MCCOY, and Martin, who led the co-location project, made several observations, paraphrased below, regarding the impact of geospatial visualization. Their observations illustrate a set of corresponding principles that apply broadly to mapping and cross-sector collaboration.

- **Observation 1:** The task force felt invigorated by the data and maps and found that studying them saved time and kept individual members from getting sidetracked by their own perspectives.
- **Principle 1:** Facts matter. An effective collaborative response requires a shared understanding of the conditions that give rise to a problem and a common point of view regarding how best to respond.
- **Observation 2:** The maps allowed task force members to locate pockets of need, indicators of risk, and assets, helping them make informed decisions.
- **Principle 2:** Place matters. Inputs and outcomes are inseparable from the context that gives rise to them.
- **Observation 3:** Mapping helped people see the data from a human perspective. Numbers and statistics became people and people's lives. Layering bus routes on top of the asset maps helped committee members grasp how difficult it is for families that rely on public transit to access services. Someone who has a car may not understand the challenge.
- **Principle 3:** Maps matter. Storytellers know that to engage an audience they must set the scene and help the reader relate to the key characters.[9] Because people intuitively situate themselves in space, maps create a bridge between reader and subject, impelling readers to project themselves into the story.
- **Observation 4:** The advisory committee saw the data and asked: How did this problem arise? Why in this neighborhood? What interventions are necessary to turn this issue around?

- **Principle 4:** Visualization matters. Humans can process larger amounts of data when it is presented visually. Patterns, trends, and relationships are easier to identify, creating a rich substrate for analysis.
- **Observation 5:** The data provided the committee with overwhelming evidence that no organization acting alone could meet the multiple complex needs of vulnerable families. Treating part of the problem would be like bandaging a gaping wound. Achieving success would require all of them to work together.
- **Principle 5:** Collaboration matters. Wicked problems are, by their very nature, best addressed on multiple fronts.

Led by MCCOY and supported by SAVI, the advisory committee chose a neighborhood, worked with residents to approve a facility, and identified 30 organizations interested in moving there. Ultimately, the group did not raise enough money to construct the facility. This outcome is not atypical. Too often, the technology and will are there, but the resources and bandwidth are not. Fortunately, in the case of Marion County's Early Intervention and Prevention Initiative, the map-based collaboration helped MCCOY build a network of partners with which it still works to reduce the number of children entering the child welfare and juvenile justice systems.

Driving collaboration with passion, purpose, patience, and data

Helping people see their stake in solving a problem collaboratively often takes a leader who can frame it compellingly. Don Brindle is one such leader. He is a longtime member of the Second Presbyterian Church in Marion County's Washington Township and a retired engineer with a passion for the church's mission in the community. In 2011, Brindle and then-minister Lewis Galloway envisioned expanding the church's Northside Mission Ministries (NMM). Second Presbyterian was set to launch a $5 million campaign to celebrate its 175th anniversary. The church planned to allocate $500,000 of that money to expand its food pantry, with more to follow as needed. But the noble idea needed thorough planning, an understanding of the factors that contribute to food insecurity, and greater clarity regarding the role of the church in the ecosystem of providers in Washington Township. Second Presbyterian agreed to rethink its preliminary pantry expansion plans.

Brindle, who volunteered to lead the effort, asked The Polis Center to conduct a street-level gap analysis to identify unmet community needs. Polis used the SAVI Community Information System to create an index of children and family needs and layered it on a map of Washington Township that also included demographic and provider information. Brindle presented the SAVI material to the planning committee of church volunteers to profound effect. By integrating diverse datasets (poverty rate, family structure, school attendance, employment, and graduation rates) and situating them in place, the needs index and the gap analyses challenged the committee's preconceived notions about poverty and resources in the community.

Washington Township is wealthier than much of the rest of the county. However, to the surprise of many on the planning committee, the maps revealed pockets of poverty. To make the data personal, Polis created one set of maps that included the home addresses of church members. This small addition had a powerful impact on church members who didn't know that they lived next door to people in poverty. What began as an examination of food insecurity morphed into a broader and deeper understanding of community need. Examples of the SAVI maps of Washington Township in figures 3.6, 3.7, and 3.8 show the food gap analyses.

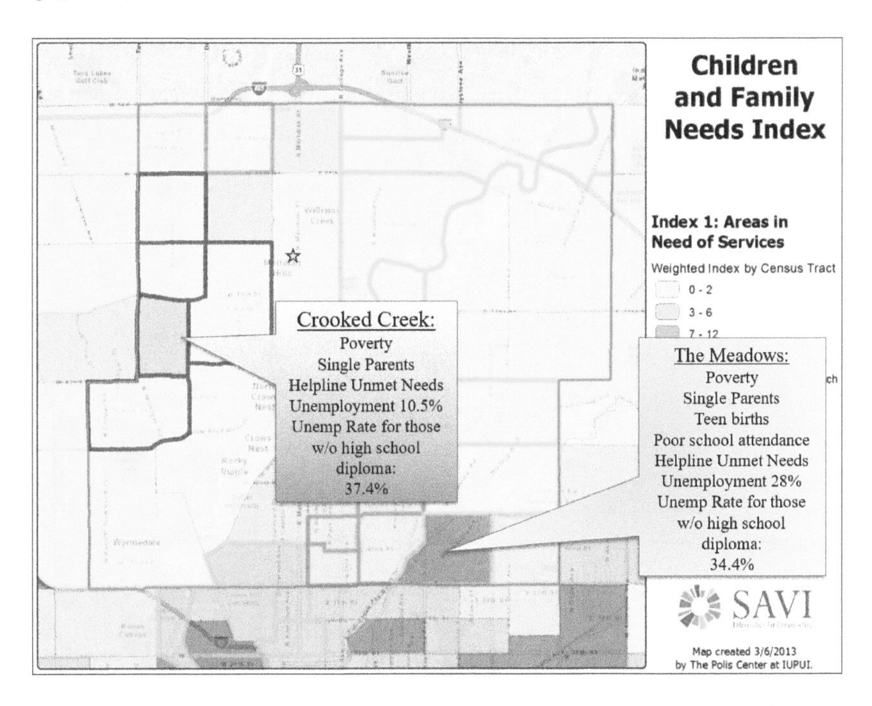

Figure 3.5. Pockets of poverty and unmet needs in Washington Township.

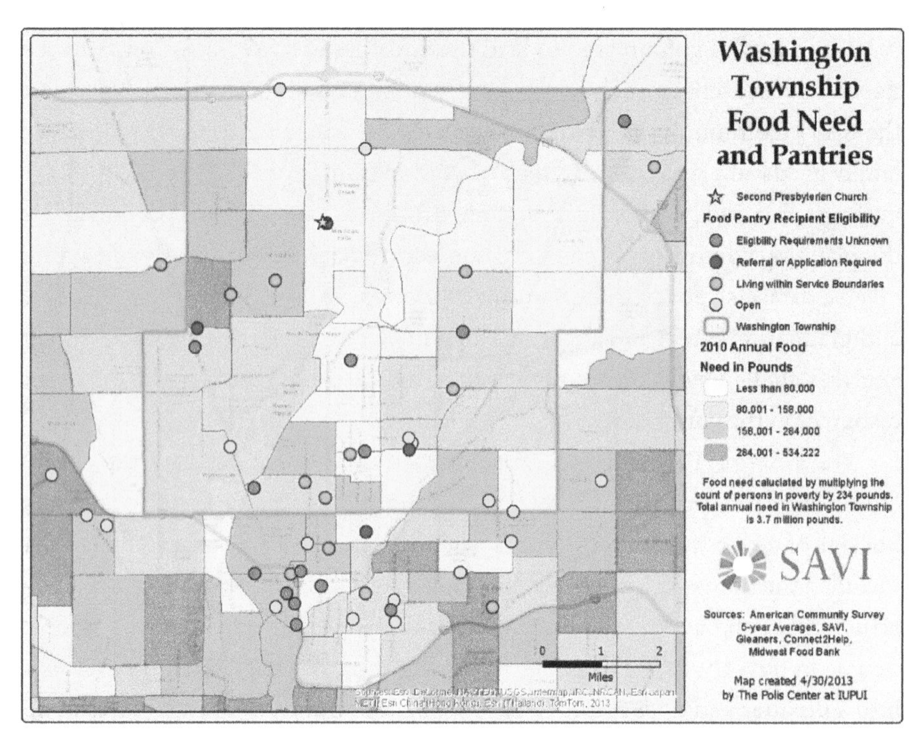

Figure 3.6

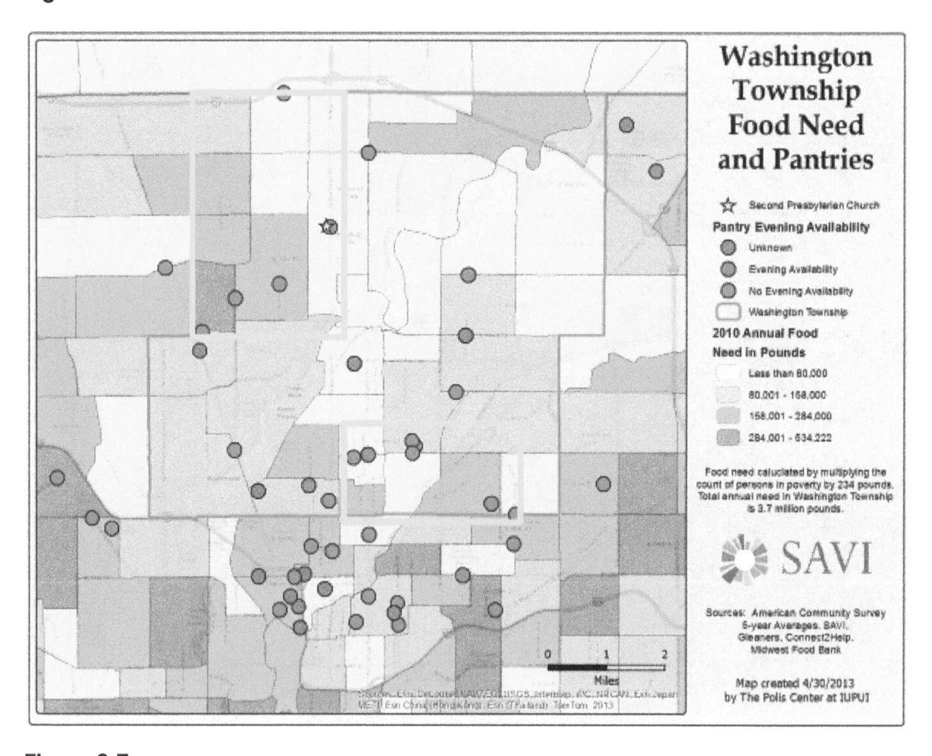

Figure 3.7

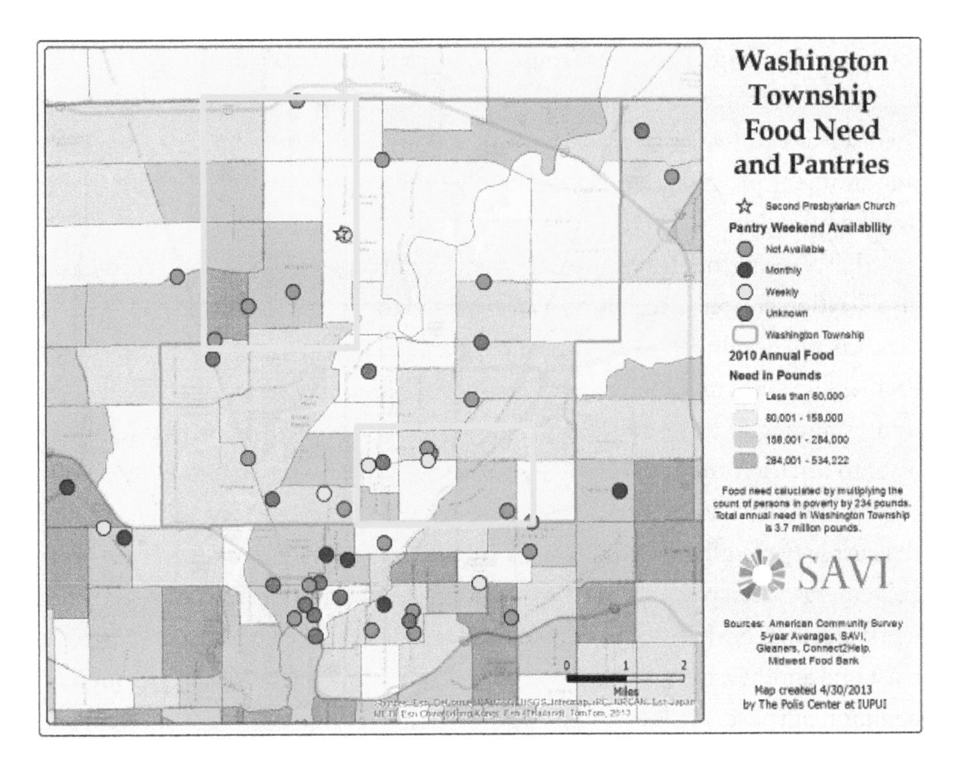

Figure 3.8

Figures 3.6, 3.7, and 3.8 plot pounds of food needed relative to food pantry availability and access.

Locating the need geospatially showed that food insecurity was just one of many challenges affecting Washington Township residents. From the mapping exercise, SAVI determined that the community's top unmet needs included housing, utilities, holiday assistance, food, and clothing. Follow-up interviews with residents revealed a lack of coordination among providers. The committee recognized that the church would have to find partners if it wanted to more fully respond to the needs of its neighbors.

The committee presented the data to the full congregation and then to the community. NMM invited other churches, food pantries, human service organizations, and representatives from county and city government and the school system to participate in the review. NMM used the maps and surveys to generate conversation with local organizations about what they were currently doing in Washington Township and what they would be willing to do to advance a collective response. Ultimately, a coalition of 25 organizations came together to form the Northside Partners Group. They started in 2011 and are still operating today. Rather than establishing a formal collaborative body, the partners consider themselves a network. They share information, coordinate activities, augment one another's programs, and occasionally cede responsibility to one another to avoid duplication.

The Washington Township experience offered several lessons. Place-based visualization created a common, if unexpected frame, from which NMM moved forward and developed a shared agenda, a crucial task in forming a collaboration. Partnership takes many forms, and no one form guarantees success. Based on its understanding of community dynamics, NMM decided early on that it did not want to force a formal partnership and would rather let collective action evolve organically. This choice worked well for them. Finally, knowledge has been called the currency of collaboration. Shared knowledge generates trust, and new protocols emerge when partners pool information and debate its implications for their work together. NMM's loose confederation functions well because the partners trust one another, meet regularly to share information, and are clear about their common purpose.

Operations

In the fall of 2018, Thrive Chicago and the University of Chicago Urban Labs[10] released a report summarizing research on the city's opportunities for youth, aged 16 to 24, who are out of work and out of school. The report's authors began by observing that, despite scores of nonprofits and government agencies working hard to support them, more youth become disconnected and are at risk of joblessness every year.[11] They suggest that several issues hinder progress:

- **Needs:** Not fully understanding the characteristics and needs of this population, because the relevant data is held in multiple government and nonprofit agencies whose systems do not speak to one another.
- **Resources:** Not having a comprehensive view of the fragmented landscape of service providers.
- **Efficacy:** Inadequate evidence to know whether youth programs are effective.
- **Dissemination:** The limited extent to which proven programs have been translated into easily adoptable practices.

The report's authors based their conclusions on a study of a specific population in a specific city. Still, their observations can apply broadly, if not universally, to the social services sector. Providers and policy makers too often lack a comprehensive picture of community needs, existing resources, and program efficacy. Innovations and proven practices exist in pockets and are poorly disseminated. This fragmentation is costly in human and organizational terms. While not a panacea, collaboration, done well, addresses many adverse consequences of fragmentation, largely through coordination, more rational resource allocation, and knowledge dissemination.

This section profiles three different organizations that collaborate with cross-sector partners in formal arrangements that rely on location intelligence to understand and project need, coordinate and allocate community resources, and incorporate innovative practices into their programs.

Starting with the basics: Out-of-school time in Richmond, Virginia

In communities with concentrated poverty, participation in after-school programs is associated with improved behavior and school attendance, healthier diets, and more physical activity.[12] Participation improves academic performance and reduces the incidence of juvenile crime.[13] In Richmond, Virginia, Mayor Levar Stony looked at his city's elementary and middle school participation rates for the school year beginning in 2017 and was dismayed. Only 15 percent of elementary and middle school students participated in full-service after-school programs.[14] In August 2018, the mayor announced plans to provide high-quality, out-of-school time (OST) programs for all of Richmond's elementary and middle school students.

Earlier that year, Mayor Stony convened a cross-sector collaborative to help the Richmond Public School System (RPS) take its first steps. The group was made up of representatives from the nonprofit provider community, three foundations, RPS, the city of Richmond, and community members. Critically, the mayor invited the city's largest providers of OST programming to take part. Richmond is not a large city, with a population of about 229,000 in 2019. But until the collaboration formed, no one had a comprehensive view of OST programs accessible to the school system's students and families.

Kimberly Bridges, then a doctoral student at Harvard and now an assistant professor at Virginia Commonwealth University, worked with the collaborative.

Although Bridges characterized the collaborative's mapping exercise as rudimentary, it generated actionable insights. Beginning with elementary schools, Bridges asked each of the OST providers to furnish information on the number of children served and slots available. Bridges aggregated the children and slots by school, rather than provider, to reveal significant coverage gaps. Two schools had no programs at all, and both were on the city's south side, home to its fastest-growing, underserved Latino population. The collaborative allocated new spots and excess provider capacity to these schools. As Bridges noted, "Just putting the information on paper, having people look at where there were and were not programs, was enough to catalyze really powerful discussion and build relationships in a way that pushed the collaboration forward." Today the working group is moving aggressively on the mayor's promise to universalize access to OST programming.

Weaving GIS and collaboration into every stage of the value chain

The United Way (UW) has served Canada since the early 1900s. Today more than 80 separately incorporated UWs operate there. As in the United States, Canadian UWs raise money through workplace and individual giving campaigns and offer grants to local charities. In the past, many UWs operated as if they were fundraising arms of local nonprofits, measuring success in dollars raised rather than outcomes achieved. Not so today.

Paul Steeves joined United Way East Ontario (formerly United Way Ottawa) as senior manager of evaluation and analytics during its transformation from a federated fundraiser to a community impact organization. Geographically referenced data helped make this transformation possible. Today data is at the center of nearly everything United Way East Ontario (UWEO) does. It uses place-based data to decide which organizations and initiatives to fund as the starting point for multi-actor collaboration and advocacy efforts, and as a tool for raising donor awareness.

UWEO's commitment to location intelligence began with the Ottawa Neighbourhood Study (ONS), a pioneering effort that explores the relationship between geography and health. UWEO was a founding participant in the development of ONS, a cross-sector collaboration between academia, government, and the private and social sectors. ONS's open access mapping platform allows users to layer indicators of individual and community health onto locally defined Ottawa neighborhoods. Users select one of 10 themes (economy and employment, general demographics, education and learning, and so on) and up to two indicators for each theme. Viewers can see datasets for each indicator (such as age of dwellings, resident income, low-income prevalence, employment, and household income) at the neighborhood or city level.

Locational intelligence and collaboration are constitutive elements of UWEO's operating model. In practice, the former facilitates the latter. In a paper published in 2012, Matthew A. Koschmann and his colleagues assert that collective agency, the ability to affect outcomes as a partnership,[15] is an emergent property. It arises not simply as a function of "structural arrangements or antecedent conditions" but rather as the result of communications processes.[16] These processes involve negotiation in which "actors use symbols and make interpretations to create ... the meanings that coordinate and control activity and knowledge."[17] The process of constructing shared meaning occurs through what Koschmann et.al. call a text-conversation dialectic, which is authoritative if it successfully "attracts and marshals resources."[18] The text-conversation dialectic shapes the assumptions, norms, and aims of the collective, which in turn binds it together.

The next section describes how UWEO uses locational intelligence in each of its different roles to affect collaboration and achieve desired outcomes.

United Way as funder

UWEO funds agency partners that address community needs in three priority focus areas: All That Kids Can Be, From Poverty to Possibility, and Healthy People, Strong Communities. According to Steeves, UWEO's agency partners use data from the ONS to inform their funding requests and identify the target populations and locations they plan to serve. In turn, UWEO uses a geospatial screen to prevent redundancy and ensure that it allocates resources where they are needed most. Maps such as those shown in figures 3.9 through 3.11 demonstrate the close ties of investments to community need, in this case based on socioeconomic status.

United Way as convener and advocate

The Canadian population is aging. In Ottawa, the number of people ages 65 and older is expected to double by 2031. This increase has significant implications for the city's social and physical infrastructure and for the institutions that serve seniors.[19] In 2017 and 2018, UWEO released reports on vulnerable seniors. It partnered with organizations such as the Council on Aging of Ottawa, senior-serving

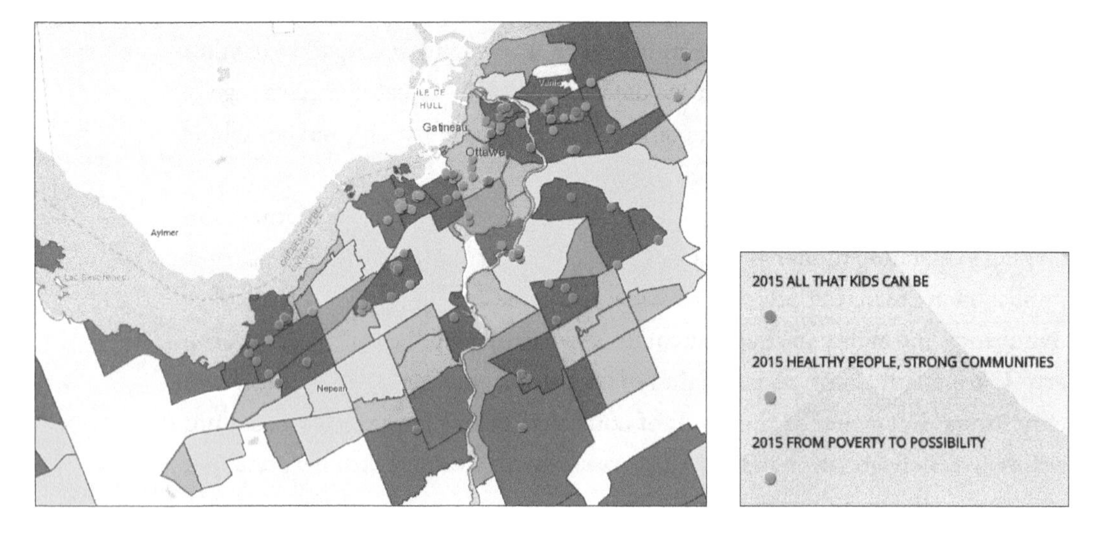

Figure 3.9. Where programs are funded by area of focus (kids, health, poverty) and community socioeconomic status.

Figure 3.10. Dollars invested by community socioeconomic status.

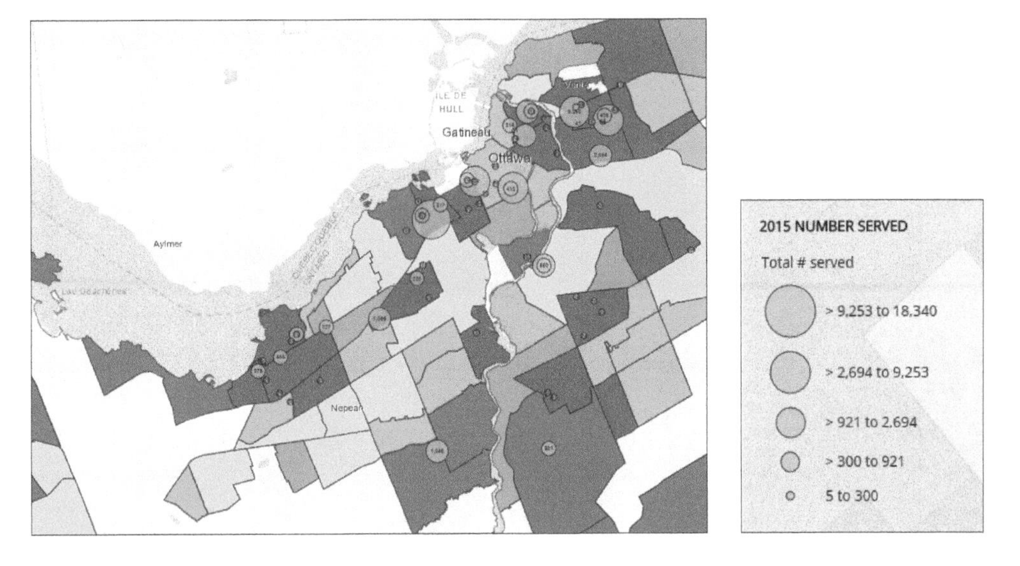

Figure 3.11. The numbers served by community socioeconomic status.

organizations, health providers, and others to understand the current and future needs of seniors in the region. The first assessment focused on Ottawa proper, and the second extended the study's reach to seniors living in rural areas. UWEO sponsored the research to better understand the future needs of the region's aging population, anticipate where to locate resources, and lobby for broad participation because no one part of the system could address all the issues.[20]

The map in figures 3.12 and 3.13 allows people to see how changes in the senior population will likely affect their community between 2010 and 2025.

By visualizing data, maps offer a geographic reference that brings people together, prompts conversations, and motivates them to find solutions, Steeves said. In the case of seniors in the United Counties of Prescott and Russell, Ontario, mapping led the cross-sector group to sign a declaration to continue to work together. The collaboration supported more nuanced thinking about the predictors of vulnerability. Working together also helped the partners focus on solutions, share resources, and align and evaluate programs. Questioning the information—and then agreeing on its meaning and implications—allowed the partners to think more deeply. This process is a necessary step in developing collective agency, in other words, the process of people coming together to get things done. Today, successful aging councils are forming across the region (in addition to the existing council in Ottawa) to meet the needs of seniors.

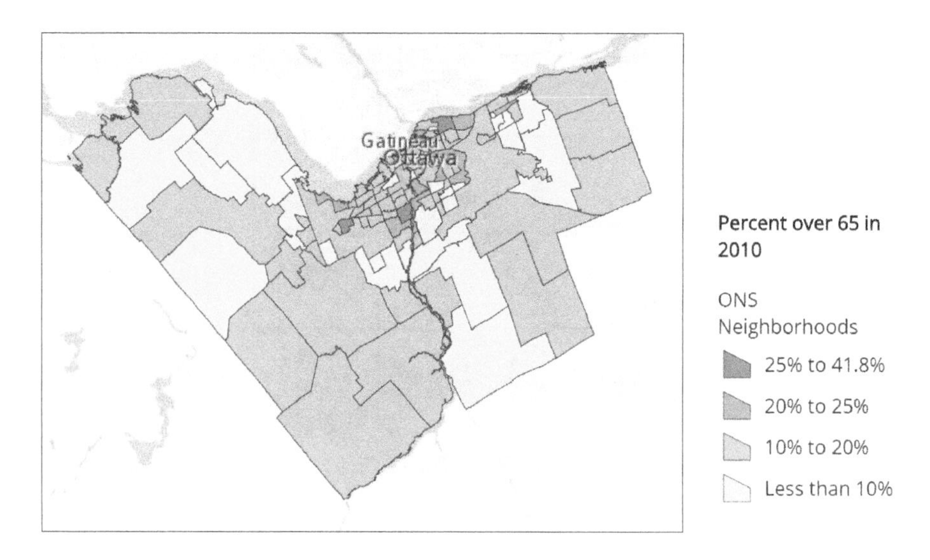

Figure 3.12. The percentage of the 65-plus population by neighborhood in 2010.

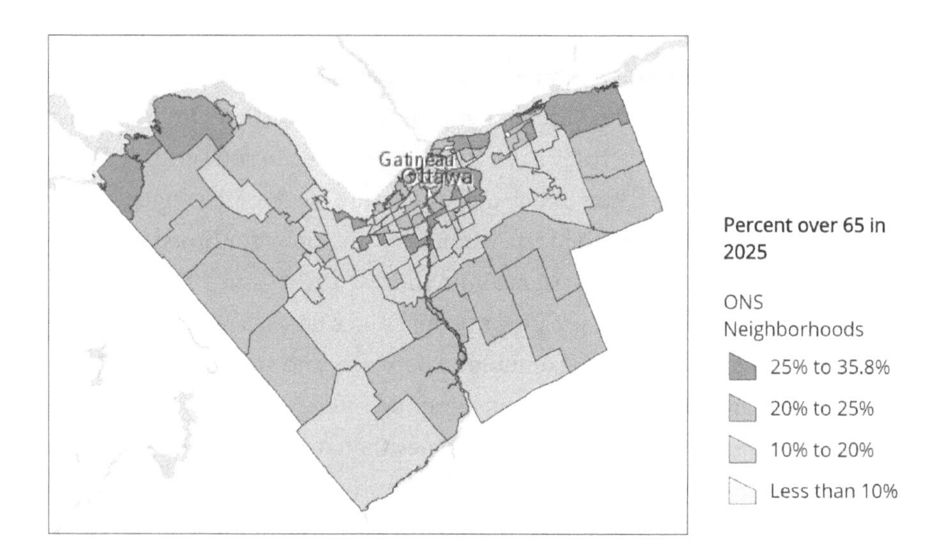

Figure 3.13. The percentage of the population over 65 projected in 2025.

United Way as fundraiser

Fundraising is a collaborative endeavor. Successful fundraisers match donors' passions with opportunities for social investment. Steeves tells the story of one donor with a passion for improving the educational outcomes of vulnerable children and youth. UWEO used geospatial analysis to determine where children were at greatest risk of getting poor school readiness scores based on the Early Development Instrument (EDI), an index that measures developmental health. EDI looks at children's health and well-being, social competence, emotional maturity, language and cognitive development, and their communication and general knowledge.

Using maps as context, UWEO and the donor strategized about where and with what agencies he should partner to address the needs of high-risk children. Figure 3.14 shows the percentage of children estimated to be at risk by neighborhood. The donor's first investment was so successful, he later funded UWEO's stewardship of a cross-sector collaboration to prevent children from losing what they learned in school during the summer break.

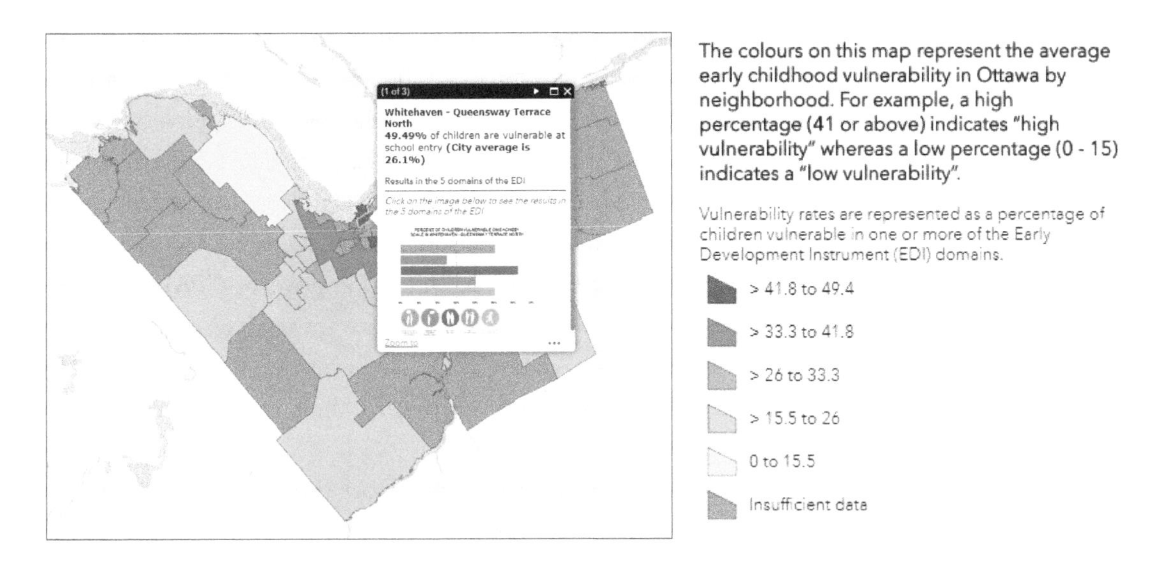

(1 of 3)
Whitehaven - Queensway Terrace North
49.49% of children are vulnerable at school entry (City average is 26.1%)
Results in the 5 domains of the EDI
Zoom to
The colours on this map represent the average early childhood vulnerability in Ottawa by neighborhood. For example, a high percentage (41 or above) indicates "high vulnerability" whereas a low percentage (0 - 15) indicates a "low vulnerability".
Vulnerability rates are represented as a percentage of children vulnerable in one or more of the Early Development Instrument (EDI) domains.
> 41.8 to 49.4
> 33.3 to 41.8
> 26 to 33.3
> 15.5 to 26
0 to 15.5
Insufficient data

Figure 3.14. The drop-down box for this map of early childhood vulnerability in Ottawa details the situation in one neighborhood for each of the five indicators that make up the EDI, including 1) health and well-being, 2) social competence, 3) emotional maturity, 4) language and cognitive development, and 5) communication and general knowledge.

Mobilizing action to fight childhood poverty

Every third child in Dallas, Texas, lives in poverty—that's more than 105,000 children out of a population of 1.3 million. Among the 10 largest cities in the United States, only Philadelphia and Houston have higher rates of childhood poverty.[21] Before he left office in June 2019, Mayor Mike Rawlings challenged city officials to address the wicked problem of childhood poverty. Meeting the mayor's challenge would require substantial cross-sector and interagency collaboration, a struggle for any city, especially one such as Dallas, where the mayor's office lacks formal authority over most municipal agencies. Like other leaders pivotal to the advancement of collaboration, Rawlings used his convening power to bring together a multi-agency, cross-sector working group of systems-level CEOs to form the Child Poverty Action Lab (CPAL). In addition to the systems-level CEOs, social and private sector representatives serve as partners and advisors.

Rawlings and CPAL's CEO Alan Cohen know the value of working across sector and organizational boundaries to solve wicked problems. Cohen said, "Our most complex problems require new levels of strategy and new levels of collaboration. That's difficult." But in the case of Dallas's children, Cohen believes that the moral and economic imperative to reduce childhood poverty will fuel greater collaboration and overcome siloed resources, historical disparity, and vested interests. Combining collective impact principles with analytics, CPAL aims to reduce the number of children living in poverty in Dallas by 50 percent in one generation.

CPAL is a good example of collaboration because it employs a disciplined, data-intensive approach to the problem of childhood poverty. CPAL uses data to inform and drive collective and individual action among its partners and in the larger Dallas stakeholder ecosystem. CPAL believes that breaking complex problems into smaller pieces is the best way to solve them through tangible and doable strategies.

Early on, CPAL exhaustively researched factors that influence childhood poverty. The partnership identified hundreds of drivers and grouped them into five categories. For each category, CPAL's partners agreed on a shared outcome, shown in table 3.1. The group also adopted six high-level strategies, called "big bets," shown in table 3.2. CPAL works with its partners and others to find, design, and implement strategies to support its shared outcomes and big bets.

To illustrate how CPAL breaks problems into smaller and smaller pieces, consider the case of access to effective contraceptives. Based on its analysis, CPAL determined that increasing the number of children living in two-parent

Table 3.1. Reducing child poverty in Dallas

Shared outcomes

North Star	Outcome categories	Target outcomes (2038)
Reduce child poverty by 50% within 20 years	Basic needs	Reduce underutilization of key family support programs (WIC, SNAP, EITC, CTC, etc.) to under 10%.
	Families	Increase the number of children growing up in two-parent homes (vs. single-parent homes) by 25%.
	Safety	Decrease high-risk Adverse Childhood Experiences (ACE) incidents by 50%.
	Education	Double the annual rate of career degrees issued.
	Living-wage jobs	Increase annual income by 25% for the bottom quintile of families.

Note: CPAL established shared outcome targets in each of five categories: basic needs, families, safety, education, and living-wage jobs to lift 52,000 Dallas children out of poverty by 2038.

households by 25 percent could materially impact the trajectories of tens of thousands of children in Dallas who might otherwise suffer from the cycle of childhood poverty.

Unplanned pregnancy increases the likelihood that a child will be born into a single-parent household. CPAL found that programs in Colorado and Tulsa, Oklahoma, that used the strategy of increasing access to contraceptives reduced teen pregnancies by more than 50 percent within a few years. By providing technical assistance to networks of health clinics and using targeted philanthropic dollars to eliminate financial barriers, the programs provided more teens with the option of choosing the most effective contraceptives for their family planning needs. Using the lessons learned in Colorado and Tulsa, CPAL is working with its partners to implement similar strategies in Dallas.

Providing access to effective contraceptives alone won't reduce generational poverty. Nor is it the only thing that affects family structure. However, implementing the strategy in combination with hundreds of other data-informed interventions is making a difference. As the next example demonstrates, CPAL uses

Table 3.2. CPAL's six big bets for action at scale

Shared strategies	
Improve service delivery for underutilized state and federal supports	Each year in Dallas County, an estimated $250M+ in existing resources for programs with a proven track record of reducing child poverty goes underutilized.
Eliminate barriers to contraception and women's health	The decision of how and when to have a family is tied to future financial security. Yet, low-income women face steep financial and other access hurdles to receiving reproductive health-care options.
Expand housing options to enable neighborhood preference	When a child grows up in a neighborhood with characteristics that are above average (vs. below average) for upward mobility, lifetime earnings for that child increase by $200,000+.
Align fragmented public investment to strengthen child care and 0–3 services	The infant/toddler years present a high-stakes opportunity to positively influence lifelong outcomes or risk lasting achievement gaps that will lessen the impact of future investment.
Reduce levels of parent and juvenile incarceration	Being jailed or losing a parent to incarceration during childhood adversely impacts future education, health, and financial outcomes. A former inmate is 7.5x less likely to see mobility from the bottom quintile to the top quintile 20 years later.
Integrate trauma prevention and care competencies in existing systems	Without intervention, trauma in childhood stunts cognitive, physical, and behavioral development. Children who experience trauma are 130% more likely to report being poor as adults.

data to catalyze action not only among its formal partners but also in the larger ecosystem of providers that affect children's well-being. Every year, Dallas leaves substantial state and federal resources unused. Take the federal government's Special Supplemental Nutrition Program for Women, Infants, and Children (WIC), which serves pregnant and postpartum women as well as children ages 0 to 5 years old (figure 3.15). Research has shown that participation in the WIC nutrition program improves birth outcomes and physical and cognitive development, which in turn improves learning and helps close opportunity gaps that prevent low-income children from achieving economic mobility as adults.[22] In Dallas, slightly less than 40 percent of eligible participants access WIC benefits, representing more than

$50 million in underused funding. CPAL believes that if more women participate in WIC, more children will benefit.

To explore why WIC participation rates were so low, CPAL used cluster analysis and mapped WIC centers, current enrollment, and eligibility. The data showed that the majority of WIC-eligible participants live outside neighborhoods where most WIC centers are located. It also showed that many WIC participants had to travel more than three miles to reach the nearest center. What might explain the mismatch between people and place? In Dallas, as elsewhere, people move, and neighborhoods change. CPAL hypothesized that WIC center leases were renewed automatically, without attention to population shifts.

Figure 3.15. This map shows WIC eligibility in Dallas County by census tract. The blue dots represent WIC facilities at the time of analysis, and the orange areas represent the density of the WIC-eligible population. The darker areas have a higher density of WIC-eligible participants.

When CPAL presented its compelling data to local WIC administrators, they agreed to reconsider their lease renewal processes. CPAL launched its WIC initiative in November 2018. By October 2019, WIC participation rates in Dallas had increased for five consecutive months.

The WIC example illustrates CPAL's dynamic model of collaboration to solve problems using data and progressive inquiry to drive action in and around Dallas.[23] WIC's willingness to act on the eligibility issue and access maps provides a case in point. Overall, CPAL blends systems-level thinking with tangible, localized milestones. CPAL has a well-defined and data-driven perspective on the aggregate drivers of childhood poverty, but it attacks those drivers incrementally, breaking them into pieces that its partners and others can address one at a time as discrete action strategies. The limited scope of these strategies makes them manageable; their aggregate impact makes them significant. As CPAL's Cohen said of the WIC action strategy, "Will we go from 40 percent participation to 100 percent? No, but this is only one of what will be a series of a dozen levers we are going to pull to address nutrition. If we try and take on the whole problem all at once, we'll never get anywhere." Finally, CPAL's approach has an emergent quality, evolving over time from data in conversation with multisector providers in Dallas.

Adaptation

Adaptation is one of the fortunate by-products of a well-functioning, data-fluent collaborative. The ongoing process of negotiating diverse points of view leads partners to question the status quo. Smart, informed discussions about what works force teams to consider alternative approaches. The consensus and trust that come from data-driven debate among partners encourage action. The result is adaptation.

Rethinking investments

It's not often that citizens vote to reinstate a tax when the economy is in dire straits. Yet this is precisely what voters did in Miami-Dade County (MDC), Florida, in 2008 when they reauthorized The Children's Trust (TCT) by a margin of nearly six to one. Florida law permits counties to create dedicated funding districts for children, provided residents vote to support them. TCT was created as one such district in 2002 through the levy of an ad valorem tax which, thanks to MDC voters, is now permanent.

TCT's governance structure is one of the more formal arrangements profiled in this book, which includes a variety of institutional arrangements under the umbrella of cross-sector collaboration. Statute determines the composition of TCT's 33-member board, ensuring collaboration that includes leaders from private and social sectors in academia and philanthropy. The board also includes representatives from the school system, courts, child welfare, juvenile justice, health department, parent-teacher-student association, universities, municipal and county agencies, state legislature, and chamber of commerce. All major systems players participate.

TCT does not provide direct services. Rather, it funds hundreds of local organizations offering a range of services that include youth development, early childhood and school-based health programs, parenting classes, and family and neighborhood supports. It also invests in program improvement and professional development for providers and staff.

Since its inception, TCT has supported early childhood programming. Most of TCT's early childhood investments finance quality-improvement initiatives and staff professional development. For more than 10 years, TCT was the primary funder of Quality Counts, an early learning quality improvement system (QIS) pioneered collaboratively within Miami-Dade County. In 2019, TCT launched a revamped QIS funding strategy, known as the Thrive by 5. Thrive by 5 has many components, but here we concentrate on TCT's adaptation of its quality improvement strategy.

That pivot began with the board, according to TCT leaders, which asked these questions:

- How do we know that our investments in quality improvement actually help children living in MDC neighborhoods?
- In what neighborhoods do families have access to highly rated programs?
- Are parents choosing these programs for their children?

The answers to these place-specific questions are best derived spatially.

TCT's board formed an early childhood working group that used observational tools to review the quality of childcare providers in MDC. TCT placed provider sites on a county map, color-coded to reflect the economic status of the children living in each zip code.

We are not showing the maps here for confidentiality reasons, but they tell a disturbing story. Despite the abundance of providers, high-quality-childcare deserts appear across the county, especially in high-need neighborhoods. For the working group, and later the full board and broader community, the maps served as a call to action.

Florida's school readiness expenditures are among the lowest in the nation. Many providers charge low rates to attract subsidy-eligible and self-pay parents who base their selection criteria on price and location rather than quality. The result is an unfortunate dynamic in the childcare marketplace. Programs that invest in quality charge more than the market can bear to cover costs. They suffer from under-enrollment, making their long-term viability tenuous. Programs that charge market or below market rates may be at or near capacity; however, they can't afford investment in quality improvement and professional development.

TCT's original quality improvement strategy did not account for these dynamics. TCT invested in quality improvement programs and professional development, but parents still didn't choose providers based on performance. TCT realized it needed a new approach. According to James Haj, president and CEO of TCT, the pivot happened when the board looked at the maps and saw quality deserts in a market over-saturated with providers.

In 2019, TCT added direct childcare payments to its portfolio. Program evaluators now use quality scores to tier early childhood providers. Providers that meet TCT's standards are eligible for "High Quality Payment Differentials"[24] of between 3 and 15 percent above Florida's School Readiness Subsidy, based on their tier, for all children attending. Child scholarships are also available for parents who do not qualify for school readiness subsidies but cannot afford the high cost of quality childcare.[25] After more than 15 years of operation, TCT adapted its early childhood investment strategy to fit the changing needs of the county. Its cross-sector board and community partners pivoted after reviewing data that presented a compelling case for change—a hard choice, made obvious by location intelligence.

Lessons for collaboration: The coordination imperative

Nowhere is coordination among providers more critical than in the social services, where hundreds of thousands of organizations labor in relative isolation, often overwhelmed by unmet needs and struggling to operate with limited resources. This chapter has profiled organizations using location intelligence to better understand community needs and allocate resources to address those needs. Yet these are only individual cases. At scale, collaboration driven by mapping and organized around place supports a more rational, less fragmented delivery system. Coordinating who provides what services in a given geography should reduce duplication, create opportunities for sharing data, and suggest where providers might benefit from combining back-office operations to free up resources for community reinvestment. With so many Americans relying on social services for support, we cannot afford to ignore the potential that collaboration offers.

Notes

1. Brice McKeever, "The Nonprofit Sector in Brief 2018: Public Charities, Giving and Volunteering," Urban Institute, November 2018, https://nccs.urban.org/publication/nonprofit-sector-brief-2018#the-nonprofit-sector-in-brief-2018-public-charites-giving-and-volunteering.
2. Institute of Medicine. *The Future of the Public's Health in the 21st Century*, (Washington, DC: The National Academies Press, 2003), https://doi.org/10.17226/10548.
3. Sally Holland and others, "Understanding Neighborhoods, Communities and Environments: New Approaches for Social Work Research," *The British Journal of Social Work*, 41, no. 4, June 2011: 689–707, https://doi.org/10.1093/bjsw/bcq123.
4. Ibid.
5. MCCOY Youth Organization, "Early Intervention and Prevention Building a Foundation for Family and Community Success: A Strategic Plan to Prevent and Reduce Child Abuse, Neglect and Delinquency in Marion County, Indiana 2010-2013," Marion County Commission on Youth, https:// MCCOYouth.org/wp-content/uploads/2011/03/604-204.EIP%20Strategic%20Plan.0510.pdf.

6. The Polis Center, "Case Study: Using SAVI in Strategic Planning and Gap Analysis," March 2013, https://www.neighborhoodindicators.org/library/catalog/using-savi-strategic-planning-and-gap-analysis.

7. Ibid.

8. Ibid.

9. Gordon Laing, "7 Reasons Why Maps are Important," Barrachd (blog). *Barrachd*, August 7, 2005, https://barrachd.co.uk/insights/blog/7-reasons-why-maps-are-important-in-data-analytics (site discontinued).

10. UChicago, "About Urban Labs," https://urbanlabs.uchicago.edu/about.

11. UChicago Urban Labs and Thrive Chicago, "Reconnecting Chicago's Youth: A Brief on Assets and Gaps," *Thrive Chicago*, Fall 2018.

12. Afterschool Alliance, *America After 3pm, Poverty 1* and *American After 3pm, Poverty 2*. Images, http://afterschoolalliance.org/imgs/AA3PM/AA3_poverty1.jpg and, http://afterschoolalliance.org/imgs/AA3PM/AA3_poverty2.jpg.

13. "Afterschool Programming," Community Foundation for a greater Richmond, https://www.cfrichmond.org/Leadership-Impact/Our-Focus/Out-of-School-Time-Programming.

14. Ibid.

15. Matthew A. Koschmann, Timothy R. Kuhn and Michael D. Pfarrer. "A Communicative Framework of Value in Cross-Sector Partnerships." The Academy of Management Review, 37, no. 3 (July 2012): 332–354. Accessed July 31, 2018, https://www.jstor.org/stable/23218092.

16. Ibid.

17. Ibid.

18. Ibid.

19. "A Profile of Vulnerable Seniors in the Ottawa Region," United Way Centraide Ottawa, https://www.unitedwayeo.ca/wp-content/uploads/2019/08/A-Profile-of-Vulnerable-Seniors-in-the-Ottawa-Region-EN-1.pdf.

20. Ibid.

21. Corbett Smith, "Here's how Dallas leaders plan to deal with the third-worst child poverty rate among major U.S. cities," *The Dallas Morning News*, November 27, 2018, https://www.dallasnews.com/news/dallas/2018/11/27/dallas-leaders-plan-115000-children-live-poverty.

22. Ibid.

23. Ibid.

24. The Children's Trust, *The Children's Trust Thrive by 5 Early Learning Quality Improvement System*, https://www.thechildrenstrust.org/content/early-learning-quality-improvement-system.

25. Ibid.

4

Improving public health

> In any disaster caused by humans or nature, maps not only support cross-sector collaboration but also inspire action. Never has this been more evident than in the global response to the coronavirus disease 2019 (COVID-19) pandemic in 2020 and beyond. Early in the pandemic, map-based dashboards played a critical role in displaying confirmed cases, lives lost, and other information, helping communities and public health officials visualize trends. See the companion website, CollaborativeCities.com, for up-to-date examples and resources on this evolving public health crisis.

For more than two centuries, maps have helped experts understand and respond to issues related to public health and communicable diseases. Medical geography first came into practice in the late 18th century when public health officials sought to understand yellow fever in New York City. Once the investigators discovered a likely causal link between yellow fever and streets near waste sites, they adjusted their prevention and treatment efforts accordingly.[1]

The 1854 London cholera outbreak offers the most frequently cited example of map-driven insights. Door-to-door canvassing and plotting data on a map revealed how the epidemic clustered around a public water pump; from this groundbreaking finding was born the modern practice of epidemiology.[2]

Today, epidemiologists and public health experts depend on mapping in their efforts to address environmental issues, improve disease management, and promote wellness. Spatial insights assist local officials with initiatives that preempt problems and provide guideposts for partnerships between governments and community groups. These insights help address the root causes of illnesses that disproportionately affect specific communities.

Many factors determine public health outcomes, including individual behaviors and decisions, environmental conditions and social systems, and the availability of care. Action to drive wellness requires a broad cross-section of partners joining together to enhance environmental conditions and produce solutions.

Cross-sector efforts to improve public health involve the same three phases: formation, operations, and adaptation. These phases incorporate the proposition that more information, organized around a place, used efficiently and adapted accordingly, will lead to better health.

This chapter examines location intelligence and health from the differing perspectives of (1) public officials, (2) leaders of community-based organizations, and (3) nonprofit officials. These leaders used spatial analytics to form and advance collaborations that ultimately improved public health outcomes.

Collaborations for healthier communities

This chapter first looks at how one city, Chicago, built an open GIS infrastructure to facilitate collaborative action based on the spatial nature of certain health conditions, outbreaks, and health hazards. Chicago ranks among the nation's leading cities in connecting open data, location intelligence, analytics, and multisector public health initiatives. The *Chicago Health Atlas* interactively maps health-related information so people can see local rates of specific health conditions, as well as locations of various health resources.

The Chicago Department of Public Health (CDPH) formed a collaboration to strengthen the mapping infrastructure that would support broad action. In building out the Chicago Health Atlas platform, the department included the nonprofit Smart Chicago Collaborative (now City Tech Collaborative) and five local hospitals as partners. An article in Data-Smart City Solutions by then-research fellow Sean Thornton chronicled this collaboration, noting that a specific effort by the department's Office of Strategy and Innovation (OSI) brought together government, for-profit, nonprofit, and philanthropic parties to pursue data insights and policies that would support initiatives for a healthier city.[3] The CDPH involved residents, providers, and community organizations as it developed the Chicago Health Atlas and its Healthy Chicago collaboration that uses the atlas for planning and targeting purposes.

Dr. Jay Bhatt, previously CDPH's chief strategy and innovation officer, emphasizes that Healthy Chicago is critical for emergency management, disease tracking, reporting, information dissemination, and implementing health interventions in public health.

Nikhil Prachand, director of epidemiology at the Chicago Department of Public Health, underscores the importance of geospatial knowledge creation and what he calls "decision-support tools." Public health has been focused on new models—those concentrated on finding causes, social determinants, and links between conditions that create unhealthy communities and people. The shift to place-based epidemiology is useful at the population level and for communities that use data to understand their problems. When the collaborative provided access to open data, place-based epidemiology allowed people involved in other issues such as housing or clean water supply to gauge the health of a community on a broader scale.

Health officials possess much more data than available before, generated by digitized records, sensors, wearables, and socioeconomic data. Chicago officials recognize that this data infrastructure requires a well-visualized, publicly accessible portal. The Chicago Health Atlas offers a data framework that partners, agencies, government entities, and anyone else can interpret to bring about change, says Raed Mansour, director of innovation for CDPH. Using this approach, researchers can secure comprehensive and usable data more easily, and residents can become more informed consumers of data. With the atlas and various nonprofit groups providing locational context, individuals have the information they need to more proactively manage their own health and improve their environment.

As the Chicago example shows, data is valuable in many collaborations. An approach used in health technology innovation can also be used to support public health collaboration. Figure 4.1 depicts a continuum that begins with the collaborative visualizing of information in a GIS format. In the second step, the collaborative considers how to develop applications, which creates opportunities for residents to benefit from the information. The third step involves predictive analytics. As organizations collaborate, they must have access to information and a way of applying it to help people within the collaborative. Predictive analytics allows members of the collaborative to determine efficacy and redirect resources.

Creating value starts in the informatics section, which includes GIS tagging, social networking data, anonymous electronic health records (EHR) information, and data visualization. In the application development section, health and technology workers then create applications from the data, addressing such issues as preventing food-borne illnesses or the flu. In the predictive analytics section, experts working with health and community data will predict outbreaks and fashion preemptive solutions. Cross-sector collaborations can form around any one of the stages shown on the chart.

The city of Chicago demonstrated how this data-driven progression produced more collaboration and better health through the city's campaign against lead. Chicago's Department of Public Health and the University of Chicago's Data Science for Social Good team partnered to pinpoint the homes most likely to contain lead-based paint hazards. First, they collected information concerning lead risks to almost 90,000 children.[4] The models used for these predictions help the analysts accelerate problem identification and mitigation and included an array of spatial information such as the age of the dwelling and lead poisoning reports in the neighborhood. The progression, shown in figure 4.1, from crowdsourcing location information to predictive analytics, helped inspectors and community activists identify and mitigate lead hazards.

Figure 4.1. This chart reflects the extent of collaborative opportunities and the amount of information that collaborations can incorporate into their public health campaigns.

Formation: Public health infrastructure

When public and nonprofit leaders form cross-sector collaborations, they develop the data infrastructure that sets the foundation for future initiatives. As shown by Chicago officials in the application development stage, public health organizations can more easily help nonprofit and for-profit collaborative partners add value by simplifying application programming interfaces (APIs). Third-party application developers have other ways to secure data, subject to privacy limitations. For instance, Data2Go.nyc, created by the nonprofit Measure of America, allows users to map several New York City–related health variables, such as deaths from cancer or preterm births, against demographic indicators such as race, income, and education level. Understanding the relationship between health and demography is crucial for policy makers to serve the city's neediest populations.[5]

Information breakthroughs in public health infrastructure extend beyond data scientists and increasingly reach average people managing their personal health. New fitness apps and digital monitoring devices generate data and provide mechanisms for broad collaborations between health professionals and individuals. The collaboration includes everything from warning patients of health risks and suggesting exercise tips to creating support groups among people with similar diseases. Organizations join to promote healthier personal habits in areas afflicted by a high incidence of a health problem, for example, infant mortality or babies with low birth weight.

Public health collaborations rely not just on health-related information but also on layering, mapping, and tracking community-based data, including but not limited to indicators of social capital. Place-based relationships and the resulting social interactions that build civic connectiveness combine to support a community's well-being and act as the foundation of thriving and healthier neighborhoods.

This layered data extends beyond just health information. Chapter 2 noted the importance of social capital to community engagement efforts; those benefits extend into public health, producing environments in which children are less anxious as they play outside with their friends. Additionally, neighbors look after one another, friends volunteer more frequently, and residents more often involve themselves in community activities that affect well-being such as zoning, land use, environmental, or traffic risk initiatives. Social bonds produce healthier lifestyles and give individuals the confidence to oppose projects or activities that would degrade the environment in their community.

Cross-sector efforts to reduce risks: Klamath Falls

The Klamath Falls story began with a spatial analytics initiative that moved from formation to operation as it evolved into what became known as the Healthy Klamath Coalition. In 2014, Sky Lakes Medical Center in Klamath Falls, Oregon, hired two public health professionals from Johns Hopkins Bloomberg School of Public Health to help launch a wellness initiative that exposed the connection between poor health outcomes and the built environment. Their research led to the formation of a community-based wellness effort and quarterly meetings involving 40 to 50 organizations.

John Ritter, professor of geomatics at the Oregon Institute of Technology (Oregon Tech), documented the conditions that lead to the coalition's efforts. Because Klamath County ranked low nationally in wellness factors, Oregon Tech asked the hospital to provide health information under a privacy agreement for research purposes. The county geocoded and grouped the records by disease categories and census block groups. The county then performed hot spot, regression, and cluster analysis and created heat maps. The results revealed city pockets where certain disease rates were higher or lower.

Klamath Falls worked with Oregon Tech to launch what would become the Healthy Klamath Initiative, and students built a dashboard to provide data access.[6] Hospital, city, and university partners then developed a model to identify health concerns within Klamath Falls, using a web-based mapping system to communicate spatial variation in disease incidence.

This Healthy Klamath mapping initiative catalyzed the formation of the Healthy Klamath Coalition, which brought local partners together to formulate a plan, provide tools, and track progress. In 2016, the Klamath County Health Promotion Disease Prevention Program and the Blue Zones Project transitioned the effort into a more formal collaborative coalition. The coalition, which had previously operated on an ad hoc basis, formed around the problems demonstrated in the early mapping effort and added five field staff to run community health improvement initiatives.

Figure 4.2 shows the dozens of entities that became involved in the collaborative. Sky Lakes Medical Center funded the program, and data drove its initiatives.[7]

The collaboration focused on areas with high rates of hypertension and heart attacks and overlaid that data with poverty and walkability scores. These insights helped officials identify areas within the community that would benefit from education about healthy behaviors. Klamath Falls officials found an exceptionally high rate of diabetes in two areas, one north and one south of downtown communities that were also associated with high levels of obesity. According to Ritter, one area of town with a high incidence of obesity had a lower-age population, was near a commuter route, and was situated near a major park.

This map-driven effort prompted the council to fund several walking trails and new bike lanes primarily for areas that would benefit the most people in terms of health outcomes. These forms of active commuting aimed to reduce the incidence of obesity and diabetes within the region.[8] The GIS analysis was central to the county winning a $25,000 grant from the Robert Woods Johnson foundation in 2018.

Ritter plans to complete the analysis again in a few years to map any differences in health outcomes. The work is still in the early operation stage, producing maps that help the collaborative understand the "why" concerning the location of specific disease categories. This collaborative process adapted its original goals to include additional subjects such as locating sources of available fruits and vegetables in the targeted areas. Ritter created a smartphone app to help students in the cadre log the locations of each store and the products it carries. The hope is that the incidence of obesity and diabetes in the area will decline, thereby improving the health of residents in Klamath Falls.

Today, the Healthy Klamath Coalition continues to adapt its goals as it learns more about the community and builds on its successes. The adapted goals come with several spatial questions:

- What routes to school are safe for children?
- How can we improve public access to grocery stores?
- Where are the food deserts?
- Where is tobacco sold near schools?

Answering these questions requires continuing enhancements in location intelligence and partners organized around a spatial- and data-driven goal.

The Klamath Falls example illustrates the role of a data consortium in creating healthier communities. Next, an Indianapolis diabetes case study illustrates the use of improved spatial analytics, first to localize the problem and then to diagnose the cause of the disease and any health conditions that might arise.

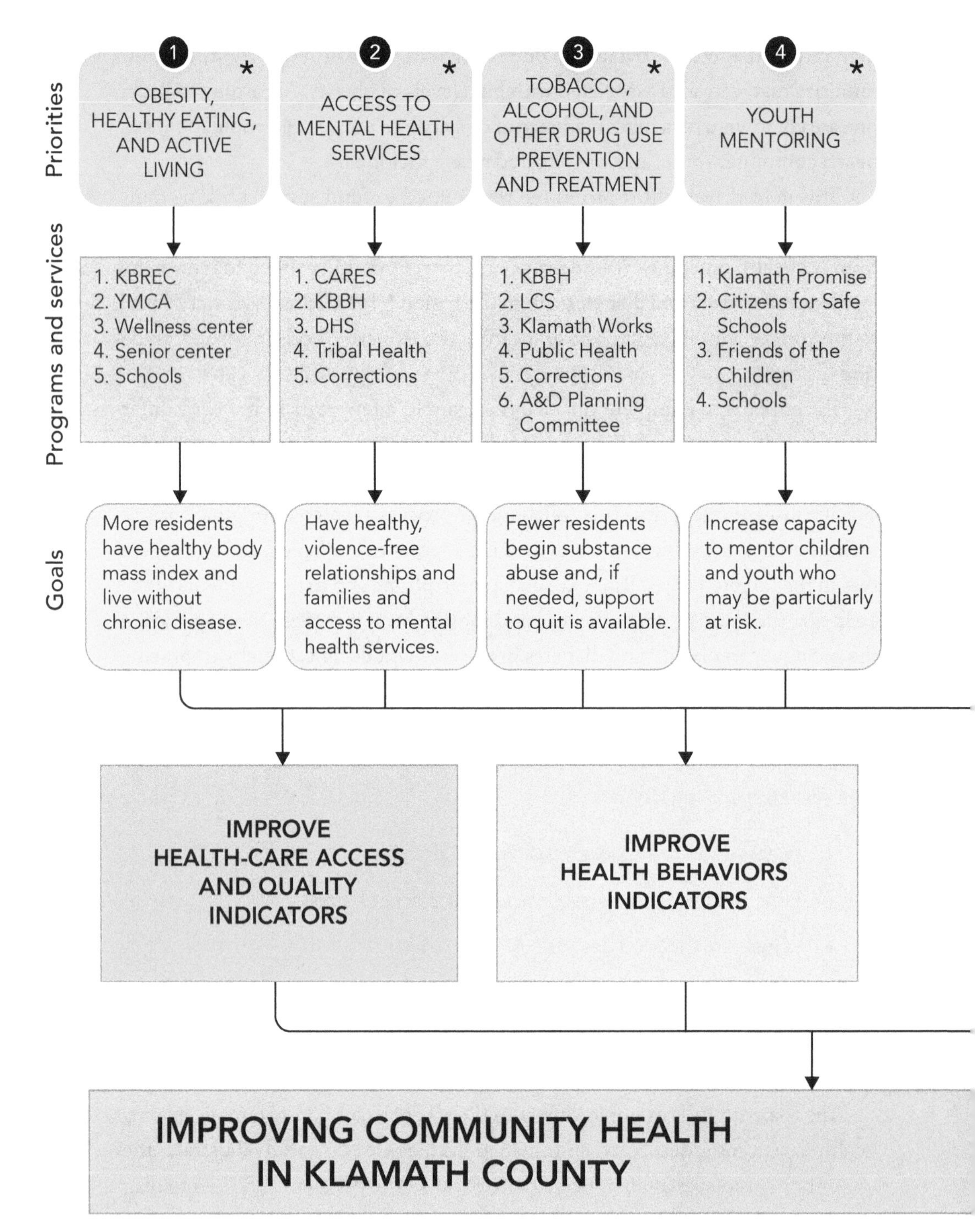
Priorities
Programs and services
Goals
1
OBESITY, HEALTHY EATING, AND ACTIVE LIVING *
2
ACCESS TO MENTAL HEALTH SERVICES *
3
TOBACCO, ALCOHOL, AND OTHER DRUG USE PREVENTION AND TREATMENT *
4
YOUTH MENTORING *
1. KBREC
2. YMCA
3. Wellness center
4. Senior center
5. Schools
1. CARES
2. KBBH
3. DHS
4. Tribal Health
5. Corrections
1. KBBH
2. LCS
3. Klamath Works
4. Public Health
5. Corrections
6. A&D Planning Committee
1. Klamath Promise
2. Citizens for Safe Schools
3. Friends of the Children
4. Schools
More residents have healthy body mass index and live without chronic disease.
Have healthy, violence-free relationships and families and access to mental health services.
Fewer residents begin substance abuse and, if needed, support to quit is available.
Increase capacity to mentor children and youth who may be particularly at risk.
IMPROVE HEALTH-CARE ACCESS AND QUALITY INDICATORS
IMPROVE HEALTH BEHAVIORS INDICATORS
IMPROVING COMMUNITY HEALTH IN KLAMATH COUNTY

* Areas that recreation has shown to improve either directly or indirectly. Associations not causality.

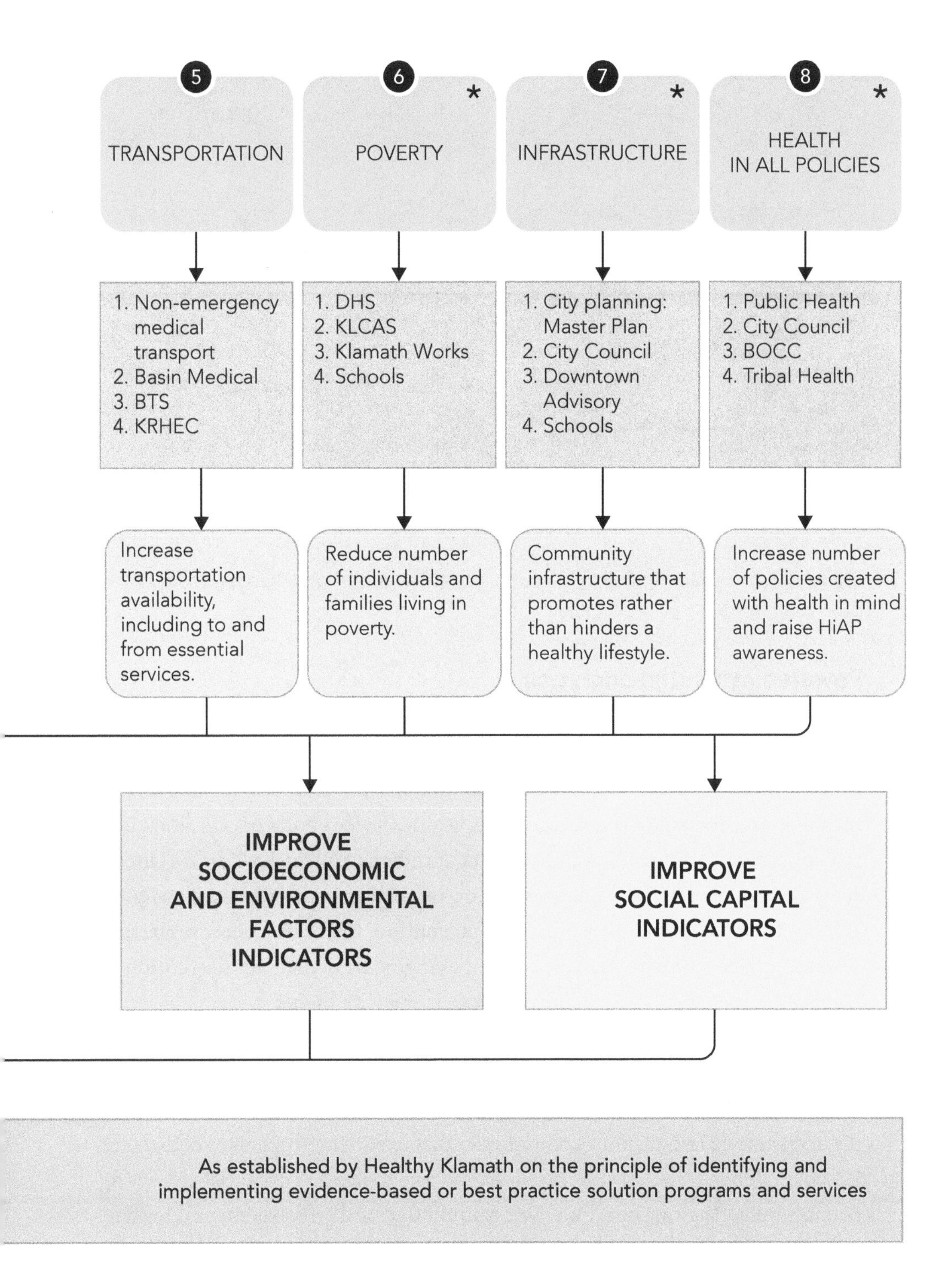

Figure 4.2. This figure (previously featured on the Healthy Klamath website) highlights the four pillars of Klamath County's Community Health Improve Plan and the goals, supporting programs, and priority health problems to address under each pillar.

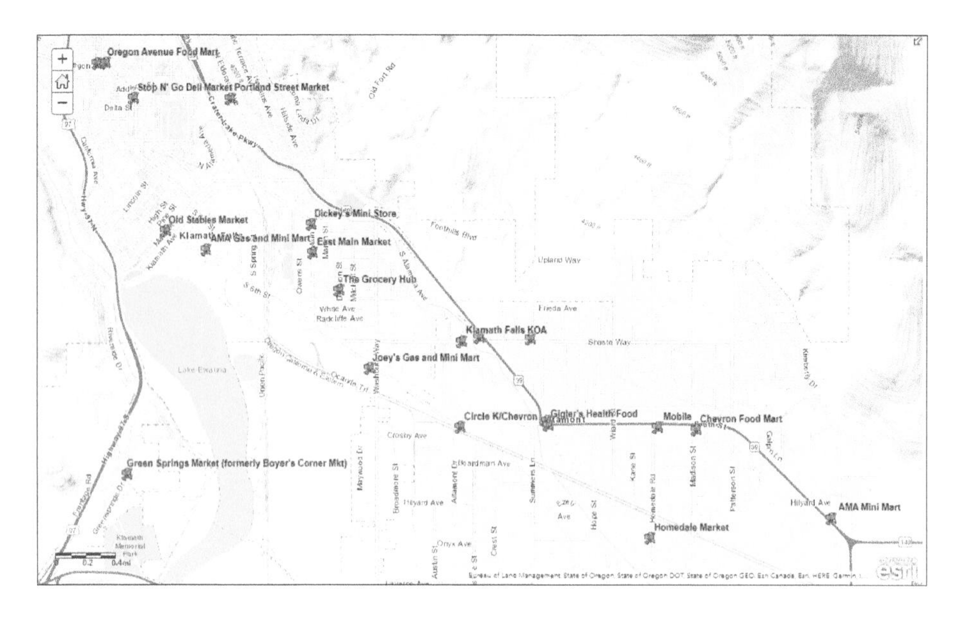

Figure 4.3. This map highlights Klamath County neighborhood stores denoted by green cart icons.

Powered by spatial analytics

Government agencies and nonprofits often lack enough resources to resolve serious problems without additional support. Adequately addressing the manifestation, say asthma, instead of the cause—say air quality—is especially difficult. But mapping data about risk factors can focus health service efforts on a location and the underlying problem. The Polis Center at Indiana University-Purdue University Indianapolis (IUPUI) has worked for decades to improve spatial analytics skills in Indiana. In a study on diabetes prevention, the Polis Center supported a broad collaboration focused on activities ranging from diabetes prevention to helping improve the quality of life of people living with diabetes.

The Regenstrief Institute, in partnership with the Polis Center, geocoded clinical records for several years to produce spatially enabled health data. Researchers mapped poverty and other risk factors associated with diabetes to illustrate the diversity among Indianapolis communities that experience high rates of diabetes. School of Public Health researchers used the data to identify potential partnering communities. They approached the communities, and cross-sector and residential collaborations formed in each of the three Indianapolis communities with high rates of diabetes.

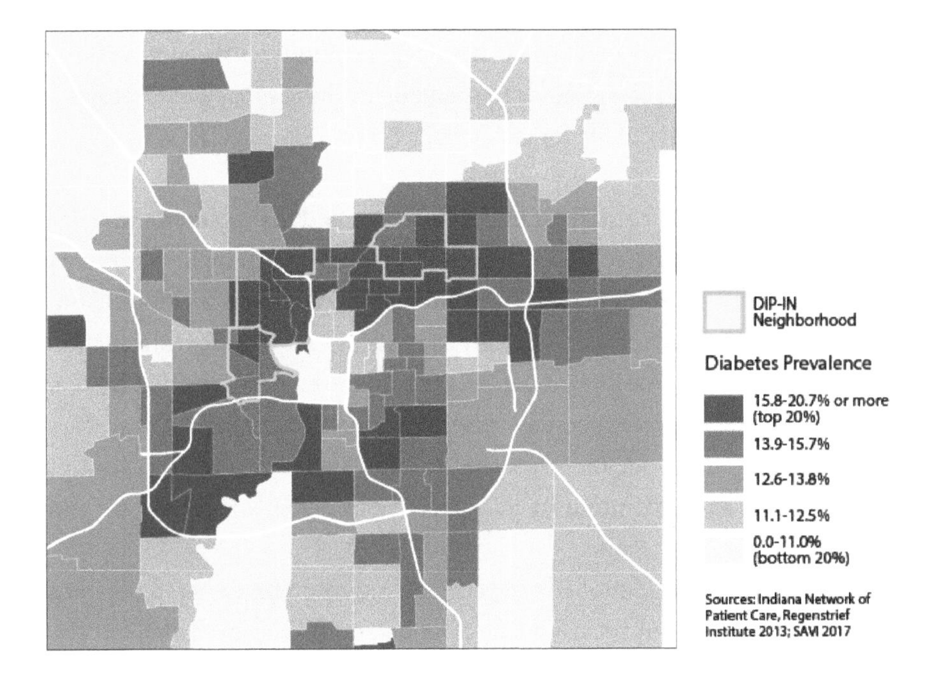

Figure 4.4. Collaborations that formed in these three Indiana neighborhoods, framed in orange, used social assets and vulnerability indicators (SAVI) to map data for improving public health in Indianapolis. This map shows diabetes prevalence of people with diabetes in each census tract in 2013 where the rate of incidence increases as the blue colors get darker.

The breadth of the spatially driven collaboration is evidenced by criteria used to create the collaborations:

- Each neighborhood will have a common thread of community health workers housed at a local public health clinic and at a neighborhood organization.
- All workers will live in the community, focus on connecting individuals to medical and social resources, and promote healthier living in their neighborhoods.
- Their duties include scheduling appointments, coordinating transportation, and seeking community access to healthy food.
- The program believes that residents can best determine the most effective resources and actions for their community. A team of residents, health workers, and program representatives will offer guidance on resources and programs they want for their community.

- Community health workers will have paid positions with benefits. The community will receive funds to spend on resources that will improve health and quality of life for residents.[9]

Karen Comer, a researcher at IUPUI in Health Policy and Management, explains that maps engage residents, so they can see different aspects of their community. The maps provide a picture of what's happening and spark conversations. Community health workers can respond quickly to residents' requests for information. Their conversations become an iterative process, not just a data presentation. For example, a map can pinpoint food deserts within a community that lacks grocery stores with nutritional food. The maps are tools for visualizing data and prompting conversations about community needs related to diabetes intervention and prevention, Comer says.

The place-based data identified the participating community-based organizations and faith-based groups serving target populations in the areas of need. The geographic analysis also identified and included multiple organizations that could provide everything from transportation to case management and medical services. As a result, collaborating organizations could better target their resources to help improve the quality of life of people living with diabetes.

Operations: Cross-sector preventive efforts

The Flint, Michigan, lead poisoning controversy not only brought to light the water quality issues that affected thousands of city residents but also raised fundamental questions about the power of information:

- What information, and in what format, should have been available to citizens?
- How could that information be delivered in a way that would help residents understand the risks in their neighborhood and allow them to act to protect themselves?

Health information organized geospatially has the power to drive individual and collective action to mitigate harm and prevent crises from happening in the first place. Compare what happened in Flint, Michigan, to the way Cerro Gordo County, Iowa, responded to its own water hazards. In 2004, county officials learned from a local resident that the shaking in her hands and neurological issues resulted from consuming high levels of arsenic in the water from her private well. In response, a collaboration formed around the goal of using spatial data to identify and reduce the number of arsenic-contaminated wells.

This effort, funded by a grant from the Centers for Disease Control and Prevention (CDC) and organized by the county public health department, included scientists and other technical experts from the University of Iowa Hydroscience & Engineering Department, the Iowa Department of Natural Resources, the State Hygienic Laboratory, the Center for Health Effects of Environmental Contamination, and the Iowa Geological Survey. This collaborative sampled, examined, mapped, and analyzed well water records and arsenic test results, which led to the discovery of an arsenic zone. With that information, public and participating nonprofit officials immediately notified vulnerable property owners whose wells were likely to contain potentially dangerous levels of arsenic.

Mapping exercises today increasingly rely not just on historic data but also real-time sensor information. As the Iowa team began to more accurately pinpoint at-risk areas, it added continuous monitoring devices that included tests for arsenic and pH, dissolved oxygen, and several other components that could affect arsenic levels. Experts also analyzed rock chip samples, well depth, casing information, and aquifer elevations.

The CDC grant funded the original collaboration, which helped the county move from awareness of a problem to developing and applying spatial analytics tools. The collaboration created 3D maps to show the location of at-risk wells. The initiative included an array of experts using spatial analytics and related science to identify risk areas. Brian Hanft, the county's environmental health service manager, said the maps helped tell the story to drive policy changes. Based on geological and geographical information for certain areas, the collaboration can assess whether its new policies reduced the likelihood of new wells testing positive for arsenic. After the collaboration approved new arsenic ordinances, only 2.4 percent of the wells drilled since then have exceeded maximum contaminant levels. The collaboration used that data to explain its successes to leverage additional results.

Visualizing the data in 3D offered insights not possible in 2D. On the research side, using GIS helped the team determine which aquifer the arsenic was coming from and construct a model. On the public health side, GIS helped the team find at-risk well owners who received information about the risk of arsenic in their water and an offer for well water testing, said Sophia Walsh, the original GIS coordinator and environmental health official.[10]

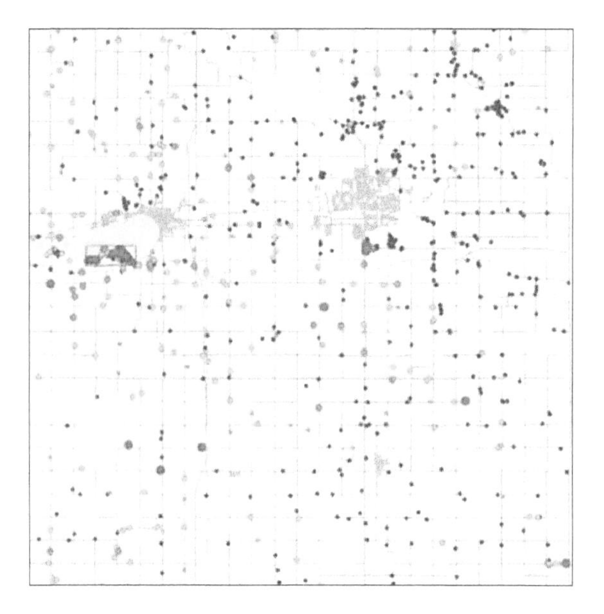

Figure 4.5. This map illustrates the levels of arsenic in private well water in Cerro Gordo County, Iowa. Levels vary from 0 to 4 parts per billion in the wells marked in green, up to 100 to 277 parts per billion in the wells marked in red. Water with arsenic levels greater than 10 parts per billion is considered unsafe.

The maps of these findings served as the foundation for a collaborative alert system. The group then adapted its operations, adding new players so that it could produce better water quality results through education and prevention. This operational adaptation included adding the local well drilling company to the collaboration, identifying the areas of risk so the company could ensure that its private customers saw the visualized data and could take steps to avoid hazardous drilling in the aquifer.

Cerro Gordo County provides an excellent example of all three stages of map-based, cross-sector collaboration. The discovery of the arsenic danger caused the immediate formation of a consortium to respond to the risk. The county Department of Public Health brought in state, university, and community partners. Further testing and map visualizations helped the group better understand the implications of drilling in certain areas. These discoveries helped the collaborative adapt its model by extending its reach to include private sector partners and incorporate education and advocacy about the location of the danger and the need to modify drilling approaches.

Monitoring, mapping, and mosquitoes

Washington, DC, health officials face a particularly challenging environment. Historically, the district's waterfront neighborhoods have suffered disproportionately from mosquito-borne illnesses, such as yellow fever, dengue, and malaria. The city also draws more than 30 million tourists from around the world annually. Of great concern is the Zika virus, which spreads when a mosquito bites an infected person and then transfers the virus when it bites another person. To defend against the spread of Zika, DC's Fight the Bite program was formed.

This collaboration, originally organized in 2016 to monitor, alert, and abate the spread of the Zika virus, also set in place a foundation that could add more partners to act in the event of an outbreak. The collaborators designed and implemented one of the most sophisticated Zika and West Nile virus monitoring and early alert systems in the United States. The effort involved a range of local and federal agencies that owned property in the area. A 2017 agreement allowed the National Gallery of Art to join the effort, giving the collaborative more information about conditions at the National Mall.

The collaboration knit together various agencies into a shared mapping effort, augmented in real time by field workers. A mobile mapping app helped inspectors collect more data, which they could instantly post to online maps with GIS tools to increase reaction times. Online and mobile tools replaced the previous manual approaches, ultimately improving timeliness and accuracy. Informed by the shared data, the partnership added more mosquito traps, distributed them more evenly throughout the area, improved the accuracy of targeted interventions, and improved readings of potential health risks in the area.[11] Through open data maps, this additional information provided better direction for mosquito control and a foundation for immediate response when needed.

Although the partnership began for mosquito monitoring purposes, the collaborative is equipped to rapidly deploy its collective assets for community outreach and early warnings in the event of a real-life Zika outbreak. In such an outbreak, community-based partners would quickly contact residents in at-risk areas about mitigation steps, such as protecting themselves with clothing and repellent. The goals and operations of the collaboration would adapt to meet changing needs.

Trey Cahill, location intelligence analyst for the District of Columbia, said the key to effective response and recovery in preventing, surveilling, and reporting vector-borne disease is geographic accuracy and spatial awareness. The public health sector increasingly uses GIS to gather, monitor, and analyze data on a map to understand and reveal relationships, patterns, and trends. Using GIS and maps to understand spatial patterns is a simple, straightforward approach to data analysis, Cahill said. Spatial patterns may emerge on a map using color gradients, differently sized symbols, and contours. GIS software can map raw data from mosquito trap or control efforts and clarify patterns quickly without the need for intermediate decisions or other analysis, he explained. Superimposing layers on basemaps with other geographic features is a qualitative, effective way to provide data to operational personnel and the public.

Spatial awareness in the public health domain allows collaborating entities to address an initial goal, which in the Washington, DC, case was monitoring. However, as the collaboration's operations mature, partners can adapt their efforts to the changing conditions with a more comprehensive reach into applicable communities.

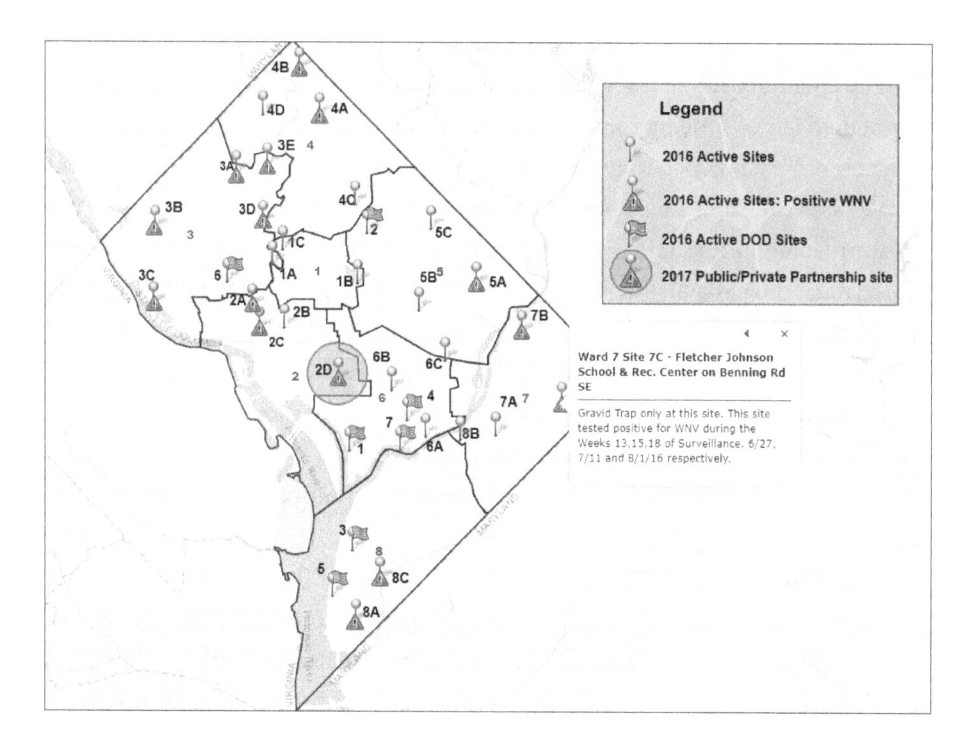

Figure 4.6. Mosquito monitoring activities in Washington, DC, include trapping and testing mosquitoes for West Nile virus. The Department of Defense managed seven trap sites, as noted by the American flag pins. In the interactive version of the map, pointing to any active site reveals a dialog box with information on that location's findings.

Zapping Zika with GIS

Zika poses a particular threat to women who, if infected during pregnancy, run the risk of their fetuses developing severe brain defects. Faced with a potential Zika outbreak, the New York City Department of Health and Mental Hygiene needed to respond quickly to identify people at risk, particularly pregnant mothers living in high-risk areas. As is typical in an emergency, a small cadre of trained professionals must act quickly, in this instance determining which of the city's 178 obstetrics and gynecology clinics should take preventive steps with their patients. A collaborative effort to reach people at risk began with the efforts of Dr. David E. Lucero of the New York City Division of Disease Control, who mapped this data to provide a risk analysis of city neighborhoods:[12]

- Population density of first-generation immigrants in NYC from Latin American countries where there is local Zika virus transmission
- Water and tree canopy density in NYC, since mosquitoes may aggregate in these areas
- Density of cases of infections caused by viruses spread by infected insects for the past three years in New York City

The department used maps to identify potential prevention partners, and city public health staff contacted elected officials, medical providers, patients, and travelers in the at-risk communities.[13] This example of a cross-sector collaborative forming and operating based on spatial data again shows how layered maps power and inform collective decision-making.

Pittsburgh air: Early warning maps

Increasingly affordable and sophisticated air sensors promise to generate enormous amounts of real-time, place-based information that will increase enforcement against polluters, as well as early warning and other alerts concerning pollution dangers. In Pittsburgh, a collaborative involving public health experts, air quality scientists, university technologists, residents, community groups, and environmental advocates formed the Breathe Project to improve local air quality.

The partnership aimed to identify areas hurt by fine particulates (PM 2.5, or particulate matter 2.5 micrometers or less in diameter) and other pollutants that form during combustion of fossil fuels, the top sources of which are industrial sites and diesel vehicles. The Breathe Project maps concentration of black carbon—soot—as a tool to help understand regional air quality.

To help produce these maps, Carnegie Mellon University researchers outfitted a mobile van with sophisticated air monitoring equipment to map concentrations of black carbon and other pollutants. Researchers drove the van around the county at different times during the year to account for seasonal variations. They paired the recorded concentrations of pollutants with a dispersion algorithm to complete a map of the entire county. Mapping showed relatively high concentrations of black carbon in the river valleys and in neighborhoods near rivers.[14]

Matthew Mehalik, executive director of the Breathe Project, explains that location intelligence drove collaborative air quality enforcement efforts and the need to construct an early warning system. The biggest public health impacts from air pollution result from fine particulate matter that humans cannot see (PM 2.5). However, people do notice bad smells, and smell serves as a proxy to eyesight for certain types of air pollutions.

The Smell Pittsburgh app serves as a sensory extension tool—because it allows people to report bad air smells to the Allegheny County Health Department, which then geocodes the reports. In effect, the health department

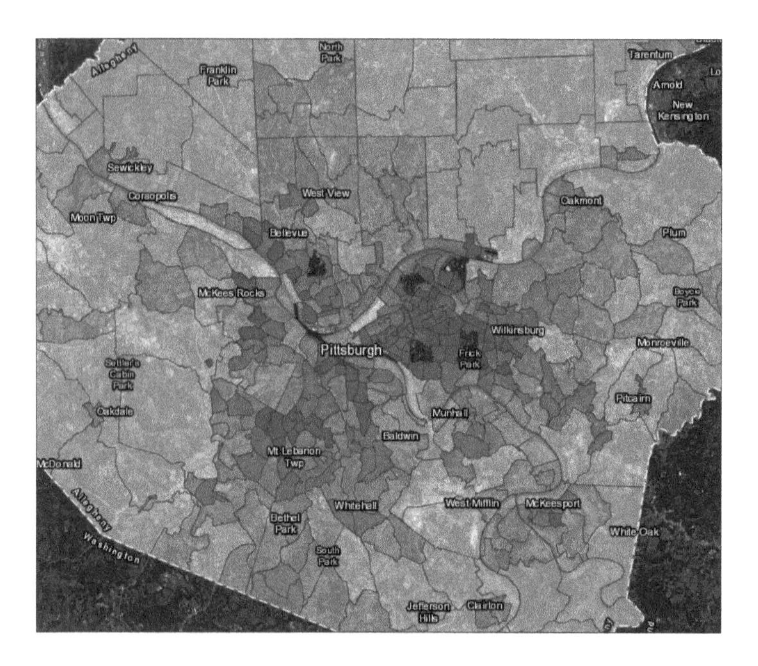

Figure 4.7. Breathing black carbon fine particles increases risks for asthma and heart attacks, among other harmful health consequences. This map showing the population density in Allegheny County allowed users to zoom in and out and click the shaded areas to obtain further information.[18] The research found elevated concentrations of black carbon in Pittsburgh's river valleys, shown in shades of red.

crowdsources smell to identify bad air across geographic regions. A large number of such reports can show how wind direction blows pollution to different geographic locations around a city so that air pollution sources can be tracked over time, which has been the case in Pittsburgh. In this way, crowdsourcing and geocoding extend a person's sensory experience of smell to track pollution and take safety precautions.

The other sensory extension tool involves sight, through the use of Breathe Cams (https://breatheproject.org/learn/breathe-cam). These high-resolution cams are trained on the region's most polluting facilities and record emission incidents around the clock. The cams sometimes record blue-gray puffs emitting from some of the facilities (as opposed to the white clouds that contain high amounts of water vapor). Researchers and regulators use the information to compare specific emission events with other data sources, such as smell reports, EPA monitoring data, and air quality index data. Residents are trained to monitor the cams and report pollution events at industrial plants to regulators and community leaders. With this information, Investigators can discover possible causes of air pollution.

Mehalik shared an example of how the collaboration worked after an accident at a coke works plant. On Christmas Eve, 2018, a fire and explosion shut down the pollution control equipment at the US Steel Clairton Coke Works, the largest coke works in the United States. Breathe Works had recorded video of the accident and shared that real-time information with state and federal regulators to ensure proper enforcement. Breathe Works made it easy for anyone to post videos or snapshots online and on social media. Concerned residents used them to prompt regulatory action. The Breathe Works collaboration offers another example of map-based data providing context for data visualization that helps residents share information, extrapolate meaning from that information, and convert that knowledge into action.

Adaptation: Informational changes and access to health services

Columbus, Ohio, officials found that access to health services involves more than simply knowing provider locations or the ability to set up an appointment. Answering questions about access requires an understanding of what factors affect timely medical visits, such as quality, medical specialty, expertise, transportation services, cleanliness, affordability, and reputation. In Columbus, residents in poorer neighborhoods often avoided nearby health-care facilities perceived to be of low quality, adding a distance challenge to their access. Researchers Timothy Hawthorne and Mei-Po Kwan of Columbus State University labeled this challenge the "satisfaction-adjusted distance measure," which they assert offers a more realistic portrayal of the issues that lower-income urban residents encounter as they attempt to access high-quality health-care facilities.[15]

The Columbus initiative adds in-depth interview results to GIS mapping to better understand quality of care and access. Within this understanding, related terms take on added meaning:

- **Availability** asks whether the supply of services can meet the population's health-care needs.
- **Accessibility** considers the location of the population in relation to the service facilities.
- **Accommodation** examines whether services at an accessible clinic meet the needs of the population.

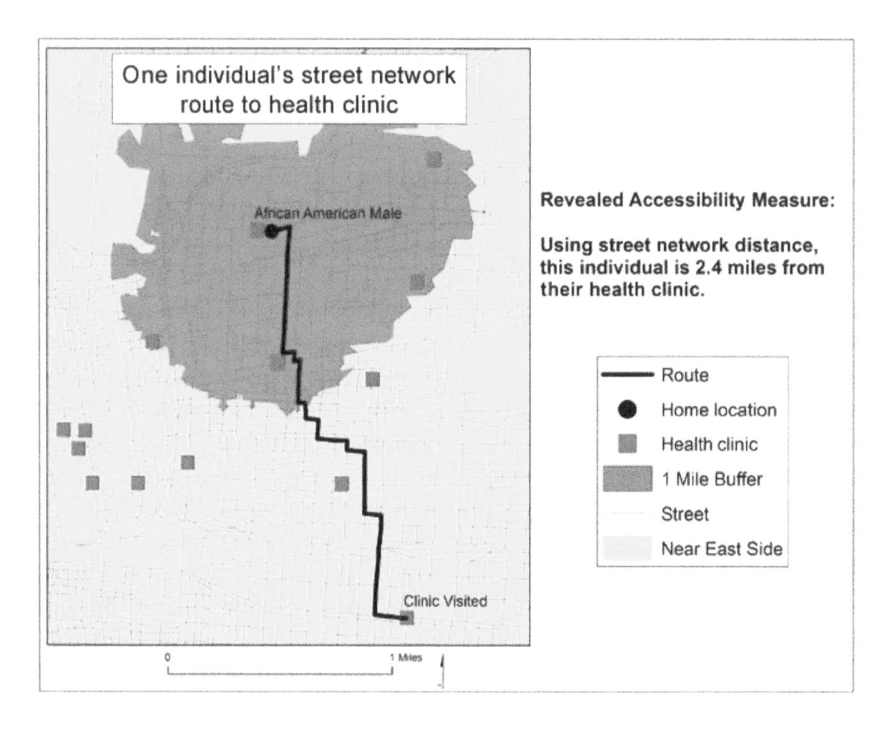

Figure 4.8. This map shows a scenario in which a lower-income resident lives within a one-mile buffer zone of five medical clinics but must travel farther to a clinic outside the zone to get quality care.

- **Affordability** considers the link between the cost of the service provider and the ability of the client to pay for the given service.
- **Acceptability** considers patient satisfaction with services provided.[16]

Although the Columbus example involves a research report instead of an actual project, the results illustrate the importance of mapping to adapt to new information. Here, one can imagine preventive health-care workers reaching a patient early in the onset of a disease. Deciding which facility to go to requires that a patient know more than just the location of a care facility that will admit them. The patient's view of provider quality and reputation will affect where they go. Impressions of a clinic sometimes result from experience and other times from perceptions. The researchers concluded that bringing better medical services to individuals requires asking the right questions and plotting the information.

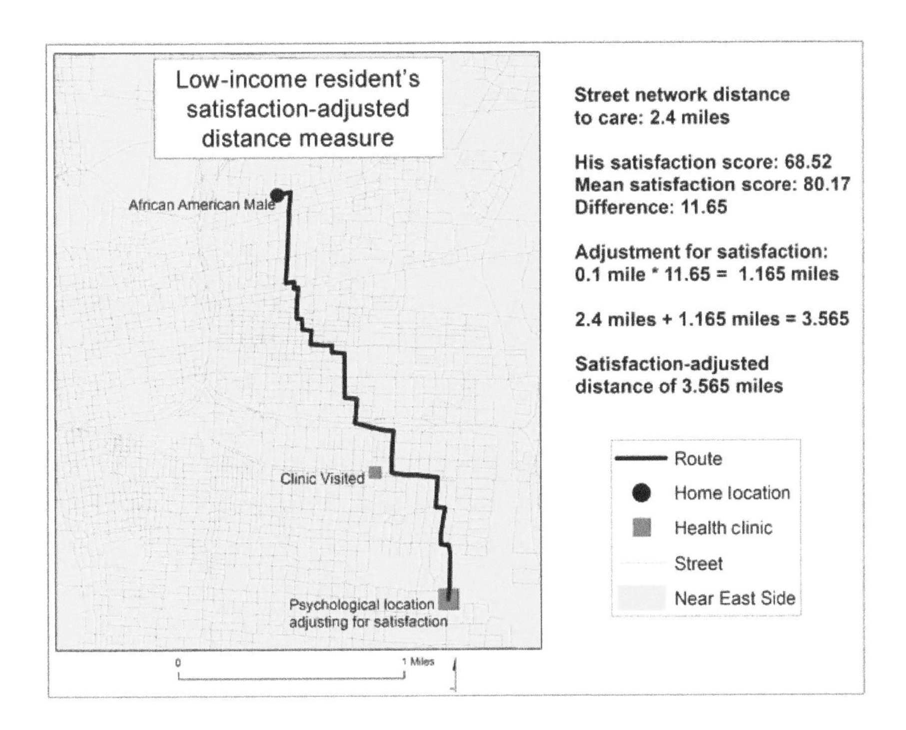

Figure 4.9. This map shows the route one person took from home to reach a medical clinic. The map is further demarcated by accounting for the quality-of-care experience the individual had. This person had an added perceived distance of 1.165 miles to reach a clinic perceived as having quality care.[19]

Chicago food enforcement

Smartphones in the hands of workers and clients, open data and sensor information, and cloud-powered GIS produce more real-time information and opportunities for collaboration and innovations in public health services. Cross-sector efforts that use these digital tools and the Internet of Things (IoT) also create an opportunity to continuously adapt and improve these services.

Public officials today operate with digital tools that allow them to collect and organize information from disparate sources. More advanced collaborations involve sharing this information through platforms. A prime example is the way food inspection has evolved in Chicago. Chicago's data efforts in recent years have featured building open data platforms with the advice of a community intermediary, the Smart Chicago Collaborative. City food inspectors generate reports when they inspect restaurants, and those reports are then made available through open data, which in turn helps inform consumers about health conditions at restaurants. This data, coupled with information from consumers through social media and evaluated through data analytics, has dramatically changed the way restaurants are evaluated.

Approximately 35 inspectors are tasked with examining more than 15,000 food establishments across Chicago. As the public began to demand better restaurant health standards, which required increased coverage and more precision, the city Health Department reached out to Chicago's Department of Innovation and Technology, a department with a reputation for advanced analytics.

The resulting collaborative included Allstate Insurance Company and the Civic Consulting Alliance, which aided the analytic work. Data came from inspections, calls to 311, and external information such as social media reports. These analysts joined to predict which food establishments were most likely to have critical violations so that they could be inspected first. In addition, researchers identified keywords in public tweets indicating a case of food poisoning. From tweets and Yelp postings, inspectors can respond via Twitter and encourage affected individuals to submit a food poisoning report.[17] As this collaboration shows, local officials with a keen sense of location and a desire to innovate can incorporate a broader group of partners in their quest for better public health.

Lessons for collaboration: Improving prevention and response

Geospatial and related digital opportunities to improve public health produce many breakthroughs. Increasingly, these real-time platforms—consisting of information from regulators, consumers, and sensors—provide early warning and precision-targeting capabilities that enable cities to improve prevention and response to disease and other public health hazards. Nonprofit and government officials using this data create cross-sector collaborations to improve health, increase access, and enhance health code enforcement. The public engages as partners in these efforts, magnifying opportunities for cities to take advantage of data and building on the promise of creating healthier communities.

Notes

1. Tom Koch, "Social epidemiology as medical geography: Back to the future," *GeoJournal* 74, no. 2, (2009): 99–106, https://www.jstor.org/stable/41148317.
2. Sean Thorton, "Data-Driven Strategy and Innovations for a Healthier Chicago," *Data-Smart City Solutions*, October 9, 2014, https://datasmart.ash.harvard.edu/news/article/data-driven-strategy-and-innovation-for-a-healthier-chicago-537.
3. Sean Thornton, "Data-Driven Strategy and Innovation for a Healthier Chicago," *Data-Smart City Solutions*, October 9, 2014, https://datasmart.ash.harvard.edu/news/article/data-driven-strategy-and-innovation-for-a-healthier-chicago-537.
4. University of Chicago, "Predictive Analytics to Prevent Lead Poisoning in Children," *Data Science for Social Good*, 2014, https://dssg.uchicago.edu/project/predictive-analytics-to-prevent-lead-poisoning-in-children.
5. Stephen Goldsmith, "The Nexus Between Data and Public Health: New analytical tools are allowing policymakers to focus on community wellness, not just on treating sickness," *Data-Smart City Solutions*, January 24, 2017, https://datasmart.ash.harvard.edu/news/article/the-nexus-between-data-and-public-health-967.
6. Oregon Institute for Technology, "Geomatics Professor Presents Research to Oregon Chapter of the American College of Physicians," January 2018, https://webadmin.oit.edu/academics/engineering-technology-management/news/2018/01/19/geomatics-professor-presents-research-to-oregon-chapter-of-the-american-college-of-physicians.

7. Healthy Klamath, "Blue Zone Project: Klamath Falls," 2016, http://www.healthyklamath.org/index.php?module=Tiles&controller=index&action=display&id=97819560063323557.

8. Klamath County GIS Department, https://kcgis.maps.arcgis.com/home/index.html.

9. Flyer, *Diabetes Impact Project, Indianapolis Neighborhoods.*

10. Sophia Walsh, "GIS Unearths the Root of Arsenic in Drinking Water: Maps Help Cerro Gordo County Department of Public Health Protect Health of Well Owners," *Esri News for Health & Human Services*, Spring 2016, https://www.esri.com/library/newsletters/health/spring-2016.pdf.

11. Trey Cahill, "Fight the Bite: Protecting the District of Columbia from Mosquitoes," DC Mosquito Story Map. https://dcgis.maps.arcgis.com/apps/MapJournal/index.html?appid=30c7550c37f04be3803c36928e6cac7c.

12. Christopher T. Lee and others, "Zika Virus Surveillance and Preparedness—New York City, 2015–2016," *Morbidity and Mortality Weekly Report*, 65 no. 24, June 24, 2016, https://www.cdc.gov/mmwr/volumes/65/wr/mm6524e3.htm?s_cid=mm6524e3_w.

13. Esri, "Zapping Zika with GIS," https://www.esri.com/content/dam/esrisites/en-us/media/newsletter/g177951-statelocal-newsletter-fall-164945.pdf.

14. Carnegie Mellon University; Center for Atmospheric Particle Studies, *The Breathe Project: Pollution Map*, https://breatheproject.org/pollution-map.

15. Timothy L. Hawthorne and Mei-Po Kwan, "Using GIS and perceived distance to understand the unequal geographies of healthcare in lower-income urban neighbourhoods," *The Geographical Journal* 178, no. 1 (March 2012): 18–30, https://www.jstor.org/stable/41475787.

16. Ibid. 19

17. Sean Thornton, "Data-Driven Strategy and Innovation for a Healthier Chicago," https://datasmart.ash.harvard.edu/news/article/data-driven-strategy-and-innovation-for-a-healthier-chicago-537.

18. Carnegie Mellon University; Center for Atmospheric Particle Studies, "Pittsburgh-Health Impact of Black Carbon Air Pollution," http://csurams.maps.arcgis.com/apps/MapSeries/index.html?appid=9b9c8dfoa6f34e3b9afcd5715872a2a3.

19. Ibid.

5

Addressing homelessness

The problem of homelessness brings together nearly all the themes of this book in their most literal form. Home is about place, place is about geography, and geography is about context. As with many social issues, the "why" often depends upon the "where." People experience homelessness for a variety of reasons. Some of the reasons have to do with economic opportunity, and the availability of affordable housing. Others are individual and concern physical health, mental health, or personal circumstances. We must respond by literally meeting people where they are and then fashioning contextually appropriate strategies. No one solution applies to every unsheltered person, in every location, across the country. The complex and interlocking issues that give rise to homelessness require many collaborative partners to respond, in many different ways. Collaborators must use location intelligence to wrangle the information and coordinate the responses to be effective.

Homelessness in America

The Homeless Hub, developed by the Canadian Observatory on Homelessness, has developed a framework for thinking about the factors contributing to homelessness, dividing them into three categories:[1]

1. **Structural factors** include the economic and societal issues that affect opportunities and social environments for individuals. These issues include inadequate income, lack of access to affordable housing and health-care, discrimination, declining intergenerational mobility, rising rents, gentrification, and insufficient education and social welfare for people experiencing acute needs.
2. The second category contributing to homelessness is **institutional failure**—Examples include difficult transitions from welfare programs and inadequate discharge planning for people leaving hospitals, corrections, and mental health and addictions facilities.[2]
3. The third group of factors contributing to homelessness is **individual and relational**, which includes an individual's personal circumstances.[3] Examples include mental health, substance abuse, domestic violence, and crises (job loss, unexpected bills, eviction).

The number of people living in official poverty in the United States increased from 23 million in the early 1970s to 47 million in 2014.[4] During this period, the concentration of income and wealth increased (rising income inequality) while access to affordable housing declined. In 1970, most renters paid less than 30 percent of their income for rent.[5] According to Princeton's Eviction Lab, "Today, most poor renting families spend at least half of their income on housing costs, with one in four of those families spending more than 70 percent of their income just on rent and utilities."[6] One of the ironies of the explosive growth in wealth seen in some cities is that it contributes to the growth in the population of people experiencing homelessness. In a study of affordable housing in Seattle, McKinsey & Co. noted, "The dwindling availability of affordable housing reflects the dynamics of the construction industry. When economic growth is strong, housing developers tend to build more profitable, expensive homes. As a result, expensive homes have become a larger percentage of the available supply."[7]

A classic wicked problem; an interconnected cross-sector response

Homelessness is often considered a wicked problem because its contributing factors are complex, interconnected, and dynamic—they play out differently depending on context and individual circumstance. What might be a plausible solution for one person in one place and set of circumstances may not be effective for another person in a different place with a different set of circumstances.

Homelessness must be addressed through the collaboration of service providers and policy makers across organizations, agencies, and sectors. Efforts to reduce homelessness commonly feature interagency and cross-sector collaboration and include these and other strategies:

- Invest in emergency, permanent supportive, and affordable housing.
- Get people into housing as quickly as possible.
- Provide support services such as case management, health care, mental health and substance abuse services, job training, childcare, legal and trauma counseling, and violence prevention.
- Improve transitional planning from custodial and health systems.
- Coordinate intake procedures, standardizing assessment protocols and prioritizing access to system resources, regardless of entry point.
- Implement tax and economic policies that support job creation, work, and the availability of affordable housing.

One chapter in a book cannot address all the complexities of homelessness affecting cities large and small, from New York City to Los Angeles and from Rockland, Maine, to Crescent City, California. This printed chapter instead looks at US cities partnering with county agencies and nonprofit providers to address the challenge of homelessness. These cities use the tools of GIS and geocoded data to advocate cooperation, facilitate an understanding of the needs of different populations experiencing homelessness, and coordinate efforts to house unsheltered people and keep them housed. They also use location intelligence to identify where individuals and families are at greatest risk of experiencing homelessness.

For a deeper dive into the wicked problem of homelessness, including case studies involving the recent complexity that the global COVID-19 pandemic has introduced, visit CollaborativeCities.com.

Formation: Making the case for collaboration in San Jose

San Jose is a study in contrasts: despite its wealth, an estimated 30 percent of its population lives below the self-sufficiency standard.[8] The standard, which varies by family composition and county within the state of California, measures how much income a family must have to meet its basic needs, such as housing, food, transportation, childcare, and taxes.

In 2016, the nonprofit Health Trust in San Jose released a detailed analysis of food access issues affecting unsheltered people and low-income seniors in San Jose. The mission of the trust is to build health equity with a focus on vulnerable populations. The assessment, titled *Food for Everyone* and commissioned by the city's Housing Department, validates the need to include food sufficiency strategies in the community's response to homelessness.

To make the case for collaboration and provide guidance to potential collaborators, analysts for the trust needed to answer several basic questions:

- Where are the city's safety net food providers (free meal and grocery programs) located?
- Where do unsheltered and recently housed people most need additional food resources?
- Do providers (safety net) have the capacity to meet the needs of these populations, and are these providers accessible via public transportation?
- Are sufficient healthy retail food options available and accessible?

To answer these questions, analysts first created maps showing the number of unsheltered and recently housed people by census tract. They added an array of data layers to the maps, including the locations of shelters, food programs (congregate, brown bag, and pantry sites), retail food outlets (including vendors accepting the federal Supplemental Nutrition Assistance Program [SNAP]), transit routes and stops, and areas within walking distance of these providers.

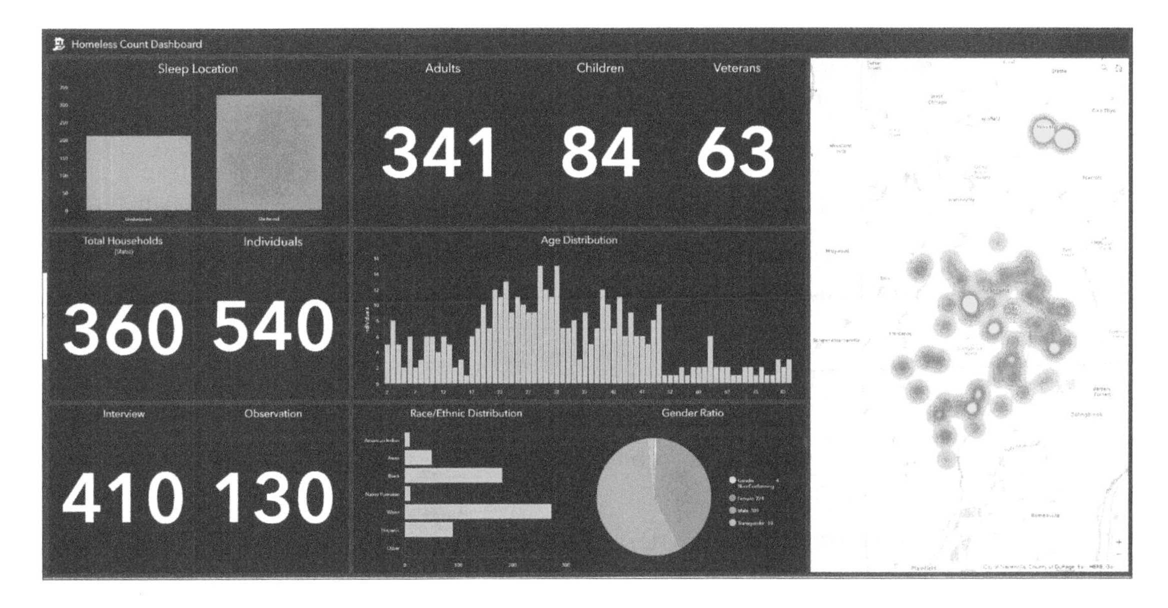

Figure 5.1. GIS-based dashboards can be used by community leaders to monitor field counts of sheltered and unsheltered persons experiencing homelessness.

Outside of downtown San Jose, the maps highlight a disturbing lack of resources. The absence of providers and access to those providers is most pronounced south of the downtown area.

The Health Trust's analysis and recommendations illustrate the need for collaboration—or more particularly, argue that existing and new community initiatives should include strategies for food access. To substantiate and contextualize its case, the trust's analysis highlights gaps in the food assistance landscape, using data regarding capacity, transportation, food quality assessments, and the availability of CalFresh (California's version of SNAP) options. The analysis is grounded in geography and represented spatially because the problems and the solutions are grounded in geography.

Since releasing *Food for Everyone,* Health Trust has organized several cooperative ventures that include formal and informal arrangements; pilot programs; and awareness, outreach, and advocacy campaigns. *Food for Everyone* turned well-visualized location data into effective advocacy that spawned partnerships with local government agencies, developers, food banks, restaurants, and other nonprofits.

Operations: Using location intelligence to coordinate and allocate resources

Homelessness challenges municipalities on multiple, interlocking fronts. Some of these challenges are tactical in nature. Although their resolution will not end homelessness, addressing them affects the well-being of both the homeless and their sheltered neighbors. Every day, city officials must help the homeless find shelter and connect both those who are still homeless and the recently rehoused with health, mental health, and support services. City officials must also keep public spaces safe, clean, and accessible. Meeting these obligations effectively and efficiently requires street outreach and the coordination of municipal and nonprofit services.

To better mesh the efforts of the organizations and agencies who interact with the homeless, many municipalities have adopted incident command system models. These operations bring together representatives from multiple areas—social services, emergency management, public works, public health, police, streets, and sanitation—to coordinate outreach and health and safety efforts.

Shared data is the backbone of these operations. Partners must understand the spatial distribution of the problem as well as the availability of necessary resources. Using mapping tools and information collected from multiple sources such as 311 calls, tent and vehicle counts, and insights from field workers, staff daily prioritize and coordinate outreach and health and public safety activities.

Unifying street strategies in Los Angeles, California

In 2018, Los Angeles Mayor Eric Garcetti announced the formation of the Unified Homelessness Response Center (UHRC). UHRC employs incident command system principles and colocates city and county departments in a single center housed in the Emergency Management Department. UHRC staff includes representatives from many city and county agencies and is responsible for coordinating Los Angeles's response to street emergencies and homeless-related service calls.

To prioritize their actions, the UHCR team must know the locations of encampments, shelters, and pending camp cleanups. To develop outreach strategies tailored to the needs of an area, the team must understand the challenges associated with each location, the characteristics of its homeless population, and the capacity of existing community-based resources.

The city relies on data visualization to answer these questions and then to coordinate outreach and site remediation. "Maps allow us to integrate all kinds of information—information that comes to us from the public, from other parts of the system, and from different departments," says Brian Buchner, chief of Homelessness Operations and Street Strategies in the mayor's office. "Mapping has become a way of life for us, and for our partners."

Making "where" more manageable through collaboration in Anchorage, Alaska

The geographic dimensions of homelessness in Anchorage, Alaska, are forcefully captured in a *Los Angeles Times* article that describes Anchorage as "a vast municipality of 1,961 square miles crisscrossed with recreational trails ... and stretches of nature ... only slightly smaller than Delaware."[9] Conducting outreach and allocating resources to address safety and sanitation issues over a territory that size requires coordination on a geographic scale that few other cities experience. Yet a collaboration of the city and nonprofits can do so by using location intelligence and information from city residents.

To assess camp locations, Anchorage crowdsources information from the community regarding encampments through its online resident portal #ANCWorks. Community-reported data is added to other data to create a dashboard of camp locations and characteristics. This information powers the operations of the Mobile Intervention Team (MIT), a formal collaboration between Anchorage's Community Action Policing team, emergency responders (including the Fire Department), social workers, and mental health providers. MIT contacts people experiencing homelessness to build trust and facilitate sharing with other organizations that can help with shelter, treatment, and counseling.

Rosa Salazar, homelessness resource coordinator for the Municipality of Anchorage, observed that GIS has helped her city better understand trends in the homelessness community, resulting in increased social services for people experiencing homelessness.[10] She also noted that the city's use of geocoded data has allowed its cross-sector teams to operate more efficiently.

No homelessness strategy can be successful without effective street-level tactics. Location intelligence cannot solve homelessness, but it can help local leaders address many of the complex day-to-day issues. Understanding the geographic distribution of the homeless population allows cities to more effectively coordinate and allocate scarce departmental and community-based resources.

Adaptation: Prioritizing outreach in Santa Clara County

The ability to respond to changing contexts, learn from experience, incorporate new information—in short, the ability to adapt—is as important for multisector partnerships as it is for organizations and individuals. When leaders of a cross-sector collaboration use predictive and spatial analytics, they stimulate conversation among partners about what works and what doesn't, and how to direct efforts and resources. For example, in Santa Clara County, California, an affiliation of municipal agencies and nonprofit providers used predictive analytics to recalibrate aspects of their outreach and engagement processes.

Santa Clara County, in California's Silicon Valley, is among the richest counties in the United States.[11] Yet the county has had one of the highest rates of people experiencing homelessness per 10,000 residents in the state of California.[12] Despite the county's wealth, the needs of its unsheltered population exceed the county's resources, and housing of any kind is in short supply. "Instead of just throwing money at the problem in a scattershot approach, we need to invest where we're going to see the biggest gains," explains Jennifer Loving, CEO of Destination: Home.[13]

Destination: Home is a public-private partnership formed to end homelessness in Santa Clara County. Destination: Home acts as a backbone organization facilitating cross-sector, collective-impact initiatives—chief among them the regional Community Plan to End Homelessness.

In 2015, Destination: Home released *Home Not Found: The Cost of Homelessness in Silicon Valley*. *Home Not Found* is one of the most comprehensive assessments of the public cost of homelessness conducted in the United States. Santa Clara County commissioned the report, produced by the Economic Roundtable, because it wanted to identify the most vulnerable among the county's population that is experiencing homelessness, tailor intervention strategies to meet their needs, prioritize individuals for housing, and where possible, reduce the cost of providing public services.[14]

Home Not Found evaluated 25 million demographic, health, mental health, and law enforcement records of 104,000 people who had been without a place to live in Santa Clara County at some point between 2007 and 2012. The study found that the county incurred costs of $520 million annually providing health, mental health, justice, emergency, and social system public services to its unsheltered population. It also found that just 5 percent of the people experiencing homelessness accounted for 47 percent of public costs and that 2,800 people representing the county's "most persistently homeless" consumed about $83,000 in public services per person per year. This cost, they observed, is far more than the annual per capita cost of providing housing and supportive services.

The study's analysis also showed that not all people chronically experiencing homelessness are excessive consumers of public services. The average cost of providing public services to someone who is persistently without housing is $13,667 per year. For homelessness service providers who face the daily challenge of allocating scarce housing resources, the opportunity is clear: Identifying by name those individuals most likely to be long-term heavy users of public services would allow outreach workers to prioritize them for supportive housing at a net savings to the county.

Using analytics, the group that produced *Home Not Found* developed a model, the Silicon Valley Triage Tool, to predict the likely heavy users of public services based on "person-specific characteristics" such as mental health, disease comorbidity, jail security risk classification, emergency room visits, and so on.

The tool is now publicly available. The county can prioritize individuals for housing when they access public resources, such as the health, mental health, or justice systems. The county also can prioritize them for placement as part of discharge planning. Outreach workers can use the model to score people who may not be identified in the system but who exhibit the characteristics of potentially high-volume users. The advantages of targeted outreach are compelling—in financial terms, but more importantly, in human terms. The Silicon Valley Triage Tool identifies the subset of people persistently without housing most likely to suffer from mental illness, health problems, addiction, and trauma. For this group, affordable housing with comprehensive services furnishes more than a roof; it provides a chance for stability and a modicum of well-being.

Most municipalities partner with nonprofits and other government agencies to engage the people experiencing homelessness. Partners often assign shelter and housing on a first-come, first-served basis. Other times, officials allocate housing based on a combination of factors, including how long a person has been unsheltered and the person's vulnerability index, a calculation based on self-reported mental and medical health issues. As shown in *Home Not Found*, these methods of allocating limited housing resources unnecessarily burden the system in terms of public costs. As a consequence, fewer individuals get the housing and services they need. Using analytics to inform outreach strategies fundamentally alters operations as usual.

Early warning systems: Moving from intervention to prevention

The next case study looks briefly at how collaborations use spatial analytics to help prevent homelessness in the first place—a fundamental and necessary adaptation. In the next example, maps help evaluate displacement risk. These publicly available maps are directed at policy makers, government, citizens, community groups, nonprofits, and businesses. By understanding the spatial distribution of risk, stakeholders can target prevention activities more precisely, rethink land-use and zoning policy, invest in affordable housing, decide where to allocate resources to fix vacant property, and catalyze public and political support.

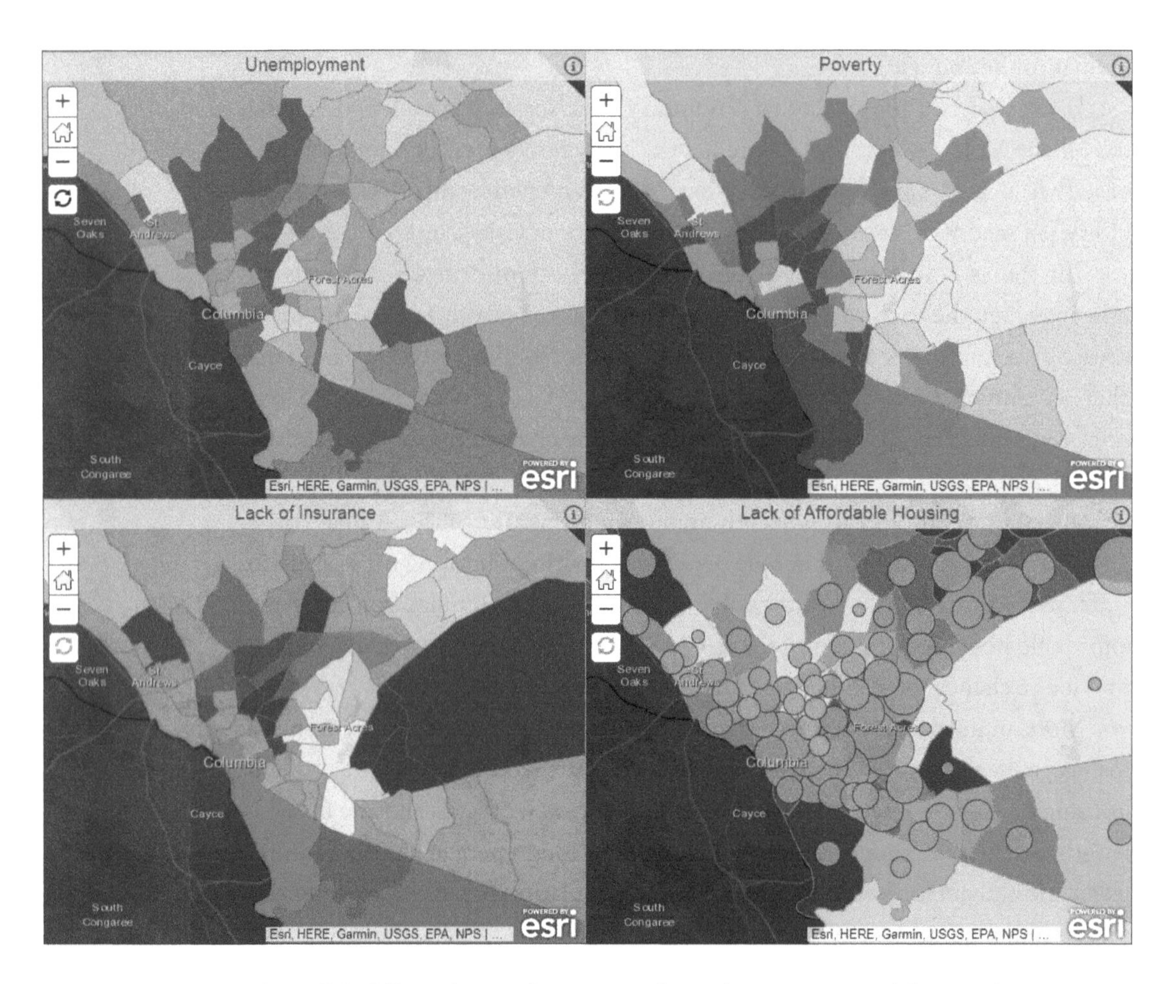

Figure 5.2. GIS can be used to map and visualize common risk factors that contribute to the current rates of homelessness, such as unemployment, poverty, lack of insurance, and lack of affordable housing.

Reducing tenant displacement in New York City

The Association for Neighborhood and Housing Development (ANHD) is a nonprofit coalition of community groups operating in New York City. In 2016, ANHD launched its Displacement Alert Project to support citywide efforts to reduce tenant displacement and preserve the city's affordable housing stock.[15] Using property level data on some 96,000 buildings in New York City, ANHD created an interactive map that estimates displacement pressure by building. By address, ANHD's analysis includes an estimate of deregulation risk (the change in the percentage of rent stabilized units), new sales risk (the most recent sale), construction risk (unique permits approved), and eviction risk (the number of marshal-executed evictions). Users also can access detailed records for each property.

ANHD provided the maps to community groups, individuals, and policymakers to equip them with information to mitigate the risk of displacement at a hyperlocal level (essentially by renter and landlord) and at a systems level (advocacy, policy, investment).

Lessons for collaboration: Addressing challenges in real time

City and county governments in collaboration with social and private sector providers strive to deliver the right interventions to the right people at the right time and in the right place. The best of these efforts relies on geocoded data to understand the spatial distribution of risk factors that contribute to homelessness. These initiatives integrate information from across systems to target and coordinate outreach and allocate scarce resources. They also adapt in real time to new data as it becomes available. For all these reasons, in their ongoing efforts to address the wicked problem of homelessness, many cities have found mapping an invaluable tool.

To continue to explore the issues of homelessness on a global scale, including the additional layer of complexity the global COVID-19 pandemic has added, visit CollaborativeCities.com.

Notes

1. Canadian Observatory on Homelessness: https://homelesshub.ca/users/homelesshub.

2. Ibid.

3. Ibid.

4. Ajay Chaudryand others, "Poverty in the United States: 50-Year Trends and Safety Net Impacts," *U.S. Department of Health and Human Services: Office of Human Services Policy Office of the Assistant Secretary for Planning and Evaluation* (March 2016): https://aspe.hhs.gov/system/files/pdf/154286/50YearTrends.pdf.

5. Bryce Covert, "The Deep, Unique American Roots of Our Affordable-Housing Crisis," *The Nation* (May 24, 2018): https://www.thenation.com/article/give-us-shelter.

6. Eviction Lab, "Why Eviction Matters," https://evictionlab.org/why-eviction-matters/#affordable-housing-crisis.

7. Maggie Stringfellow and Dilip Wagle, "The economics of homelessness in Seattle and King County," McKinsey & Company (May 2018): https://www.mckinsey.com/featured-insights/future-of-cities/the-economics-of-homelessness-in-seattle-and-king-county.

8. HealthTrust, "Food for Everyone" (Winter 2015/2016): http://healthtrust.org/wp-content/uploads/2018/10/Food-for-Everyone.pdf.

9. Hughes, Zachariah. "Anchorage has a homeless problem too, but in the woods, not on skid row." *Los Angeles Times,* August 24, 2018: https://www.latimes.com/nation/la-na-alaska-homeless-20180824-story.html

10. Salazar, Rosa. "Anchored Home: Municipality of Anchorage's Blog on Homelessness." September 18, 2018: https://medium.com/@anchoredhome/anchored-home-64720e20c0a6.

11. Rebecca Lerner, "The 10 Richest Counties In America 2017," Forbes (July 13, 2017): https://www.forbes.com/sites/rebeccalerner/2017/07/13/top-10-richest-counties-in-america-2017/#4f32024e2ef3.

12. Sacramento Steps Forward, https://sacramentostepsforward.org/2019pitcount.

13. Josh Harkinson, "Could This Algorithm Pick Which Homeless People Get Housing ?" *Mother Jones* (June 29, 2016): https://www.motherjones.com/politics/2016/06/homelessness-data-silicon-valley-prediction-santa-clara.

14. Daniel Flaming, Halil Torros, and Patrick Burns, "Home Not Found: The Cost of Homelessness in Silicon Valley," *Economic Roundtable* (May 26, 2015): https://economicrt.org/publication/home-not-found.

15. Displacement Alert Project, New York City, https://www.displacementalert.org/about.

6

Responding to disasters

When American Airlines Flight 77 flew into the Pentagon on the morning of September 11, 2001, federal and local officials scrambled to help the injured and evacuate the city. Ordered to leave with no clear directions of where to go, motorists promptly created gridlock in every intersection.

After Hurricane Katrina devastated New Orleans in 2005, residents in the lower Ninth Ward, the poorest and hardest hit part of New Orleans, complained furiously about the city's response and recovery efforts. The city's initial plans offered general ideas but few specifics to guide people who wanted to return and rebuild.

These two disasters and other examples presented in this chapter show the need for governments to respond to emergencies by focusing on place. They need to ask: Where might the catastrophe strike? How and where will they evacuate people? What infrastructure must they repair, and when? The hard lessons of experience show that location intelligence is key to the wicked problem of disaster planning and response.

When tragedies strike, maps can bring critical stakeholders together, create common understanding, and result in more effective place-based, strategic decision-making. Never are these capabilities more crucial than during disaster response efforts, when officials must coordinate across multiple organizations in real time when people's lives are at stake. This chapter illustrates the critical role that maps play in the formation, operations, and adaptation stages of collaboration before, during, and after emergencies. As the COVID-19 pandemic has demonstrated, we depend on maps and dashboards not just for collaborating across sectors but for including the public as partners in the overall response (see CollaborativeCities.com).

Disaster response, whether caused by humans or nature, requires collective action from an array of stakeholders. The growing scope and damage of a disaster limits the effectiveness of individual responders and requires intervention from state, federal, and sometimes international governmental and nongovernmental bodies. As more actors get involved, the complexity of coordinating resources and agreeing on priority actions can overwhelm even the most seasoned emergency managers. During a response, emergency managers face the crucial task of knowing where to allocate resources. An effective (or ineffective) response can mean the difference between life and death for thousands of people. Mapping can help overcome the barriers of time and complexity and serve as an effective platform for team-based collaborating and decision-making, real-time monitoring, communicating, and helping emergency responders adapt to changing environmental conditions.

Formation: Mapping collaborative prevention and preparation

A well-managed collaboration between emergency managers, law enforcement, paramedics, volunteers, and various other stakeholders serves as the basis for effective disaster response. From the start, the leaders must gather and use information about the impacted area to deploy resources where they are most needed. This work is the formation stage.

Effective emergency response depends on the quality of planning and preparation before an event. One example of these advanced preparations involves mapping environmental risks to predict how certain weather events might impact specific areas of a city. Mapping solutions can help officials work with community and neighborhood leaders to predict the most vulnerable areas. The result is better disaster preparation.

For example, National Storm Surge Hazard Maps predict storm surge levels by hurricane category nationwide. The National Oceanic and Atmospheric

Administration (NOAA) created these maps to help US communities become more resilient and better prepared for hurricanes and their subsequent storm surge, which may extend many miles inland. These maps clarify the breadth of storm surge exposure, informing outreach efforts for hurricane evacuation routes. Using these maps, emergency responders can take a range of precautionary steps depending on the type and risk of different hurricane categories. Emergency managers can also use these maps to collaborate with hospitals, nursing homes, and other at-risk locations to develop contingency plans for transportation and care.

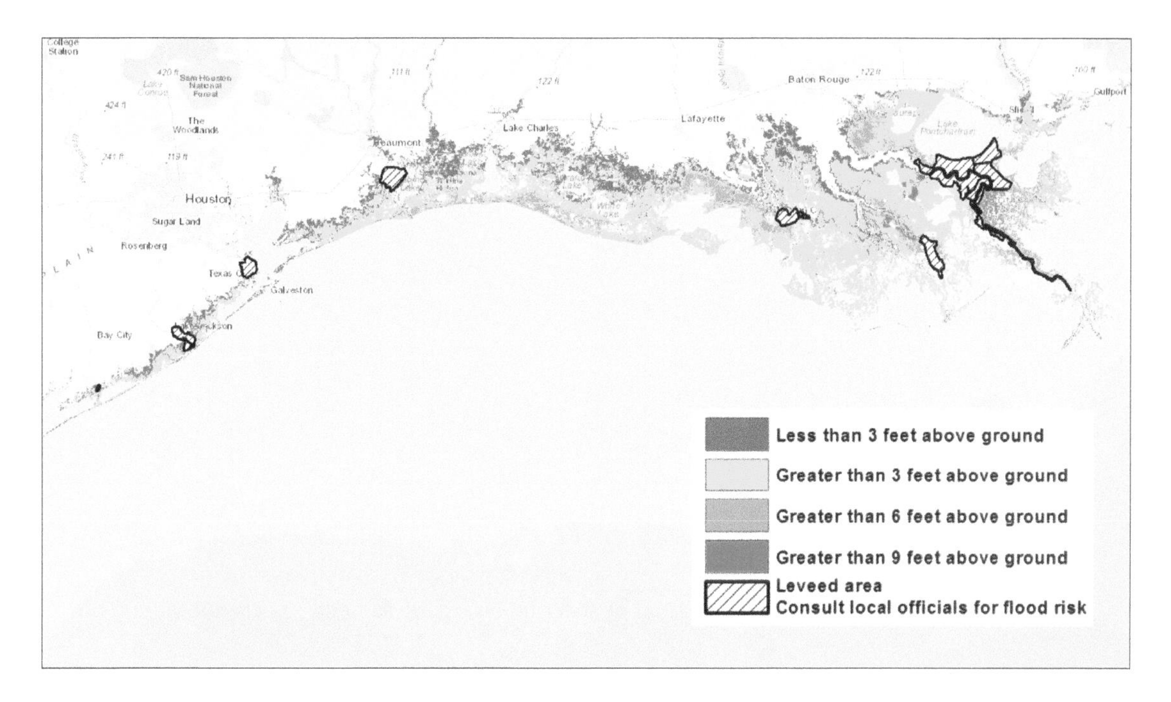

Figure 6.1. National Storm Surge Hazard Maps from NOAA highlight levels of storm surge risk along the Gulf of Mexico. This map shows the potential storm surge caused by a Category 1 hurricane, with damaging winds ranging from 74 to 95 mph.

Similarly, Seattle set up an interactive map called *Seattle Hazard Explorer* to identify specific vulnerabilities to environmental hazards including earthquakes, tsunamis, landslides, and flooding so that community stakeholders and residents could better prepare for these events. The map helps users visualize content from the more comprehensive *Seattle Hazard Identification and Vulnerability Analysis* (SHIVA) report that focuses on how 18 hazards might affect the city.

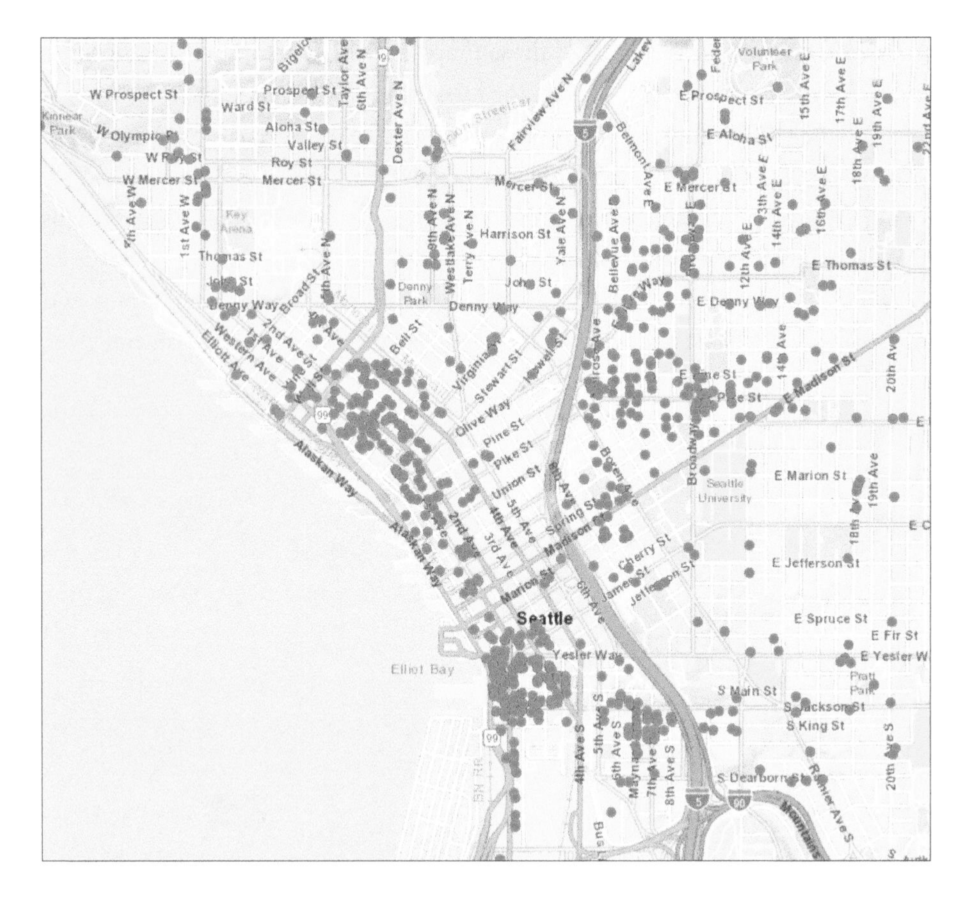

Figure 6.2. The red circles in this map identify the locations of buildings with unreinforced masonry in Seattle. Older buildings were not constructed using modern building codes and pose a greater risk of damage or collapse during an earthquake.

Elected leaders must form collaborations that use more than their own government's assets in response to an emergency; they also must engage private and nonprofit partners and community residents. Maps identifying risk can help identify who to include in the collaboration, what community they represent or can support, and how they can help mitigate a risk in a particular place. Advanced spatial planning prepares partners for the difficult operations that ensue when a disaster occurs.

The data drill: A key part of cross-sector formation

In the aftermath of Hurricane Sandy in 2012, New York City officials discovered not just physical structures in desperate need of repair but vulnerabilities in the information systems that support emergency management. New York City's Office of Emergency Management (OEM) had earned a reputation as one of the country's best from its experience and drills in preparation for emergencies. Yet James McConnell, assistant commissioner for Strategic Data for OEM, observed that Hurricane Sandy exposed vulnerabilities in the city's data infrastructure and protocols, noting that the office did not fully test its data during drills.[1]

After Hurricane Sandy, OEM and the Mayor's Office of Data Analytics (MODA) began drills to stress-test the city's data protocols in an emergency. Spatial analytics served as the foundation for these drills, which included testing each agency's ability to quickly assemble the data needed to respond to a hypothetical emergency. Responding to a disaster means responding to a place, and preparation for that response requires knowing everything possible about that place.

Amen Ra Mashariki, then-director of MODA, emphasized that it is crucial to retrieve data easily. He cited the city's response to an outbreak of Legionnaires' disease in 2015, caused by infected cooling towers. As the outbreak spread, the mayor's office needed to know as soon as possible the location of towers that might pose a public health risk. The request required MODA employees to undertake a time-consuming effort to acquire, spatially layer, and analyze various datasets to predict the vulnerable locations. The work exemplified spatial analytics in action.

During the Legionnaires' outbreak, coordinating agencies faced the challenge of getting the right data to the right people. "Right data" usually means the most current location information—where an emergency occurs and how to move the right resources to the correct place. Responders must have maps and be ready to share geospatial data quickly. Data drills allow agencies to practice sharing data and collaborate across departments so that everyone can do their jobs effectively.[2]

Learning from the Legionnaires' outbreak, city officials helped NYC prepare for emergencies by leading additional drills to tease out the needed data and identify which agencies or private or nonprofit enterprises could provide it. That data search led to an examination of data access and quality. Data drills help identify in advance which party controls the needed data and how to make that data quickly available in certain conditions. Data drills also lead to discovery of missing data or holes in the process, and help identify additional parties that should be invited to the collaboration. For example, cross-sector preparation for an emergency could include mapping hospitals, nursing homes, and shelters, including their interiors to locate generators and important stored resources.

When stress-tested in a data drill, participating agencies improved their data quality availability. The improved data represented an early success of the collaboration while laying the groundwork for future collective action. Maps and data organized during the data drill served as a foundation upon which map layers could be added, based on ground observations during an event. Data from field workers and volunteers about real-time conditions can help predict where resources must be allocated first in response to the greatest risks to public health and safety. The progress in the formation stage establishes information flows and feedback loops that become critical during the response.

Operations: Cross-sector response to emergencies

Crisis response is the ultimate test of a cross-sector collaboration. During an emergency, responders depend on communication among different teams. As information floods in from multiple sources, miscommunication and failures to communicate can lead to inaction or potentially disastrous mistakes. For example, first responders at the local, state, and federal levels often struggle to effectively communicate changing field conditions with one another once they deploy assets. By sharing real-time maps, responders can see the location of their resources in relation to changes on the ground and new emerging threats, identify communication problems, and synchronize response teams.

In a larger context, the California Office of Emergency Services uses this approach to help keep response teams informed. This real-time response capability is critically important during California wildfires. Firefighters see visualizations of the fire's movement, neighborhoods at risk, and structures threatened, damaged, or already destroyed. Individual firefighting teams can see the locations of other teams and where resources are working (or not working) to mitigate damage and then react accordingly to address the spread of the fire.

Relying on data from cameras and sensors that populate maps with the real-time movements of people and environmental conditions, emergency responders can move with precision and speed. Such real-time data can provide situational awareness in a variety of contexts. For example, the Chicago Office of Emergency Management and Communications (OEMC) monitors resources and runners

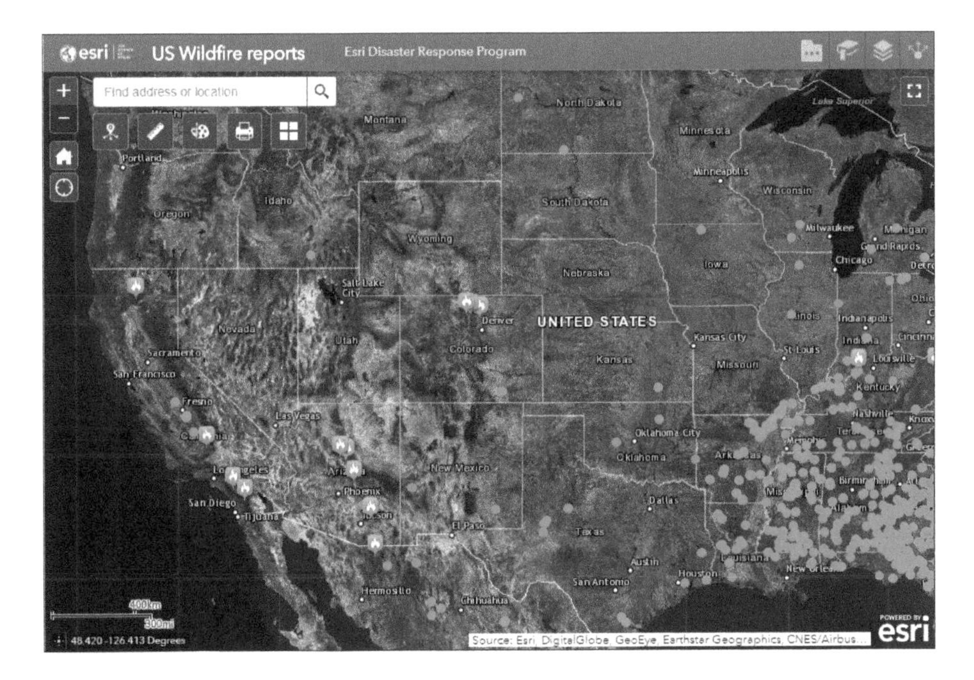

Figure 6.3. The Esri Disaster Response Program updates data used to visualize and map active wildfires or wildfire risk areas every 6, 12, or 18 hours.

at the Chicago Marathon using real-time, shared maps. Tiny shoelace-based sensors collect and send data to GIS technicians who populate maps with the real-time movements of people at the event and with environmental conditions such as temperature, humidity, and rainfall. Real-time monitoring allows OEMC to coordinate and maintain communications about race conditions, emergency response assets, and runner locations with local, state, and federal law enforcement and other stakeholders. The digital communication allows all partners to see where each runner is and how they are progressing through the race, and to quickly identify bottlenecks or abnormal movements. This location-based capability provides early warning of potential emergencies and speeds response to any race emergency, such as an injured runner or an incident on the scale of the 2013 Boston Marathon bombing.

Similarly, the Homeland Security Advisory Council (HSAC) at Pepperdine University's School of Public Policy in greater Los Angeles (LA) used maps to coordinate numerous entities involved in planning and prevention related to the 2019 LA Marathon. The LA Marathon crosses five jurisdictions, which makes information sharing and situational awareness challenging. With HSAC's SALUS app, the five jurisdictions share information in real time and maintain complete situational awareness throughout the event using a common mapping platform and dashboard.

Clearly visualizing the location of risks and assets in advance of emergencies provides the platform for real-time data sharing from sensors and cameras attached to people or objects. The quality of location data acquired and organized in the formation phase will radiate through the collaborative, improving the quality of planning and eventually, emergency response.

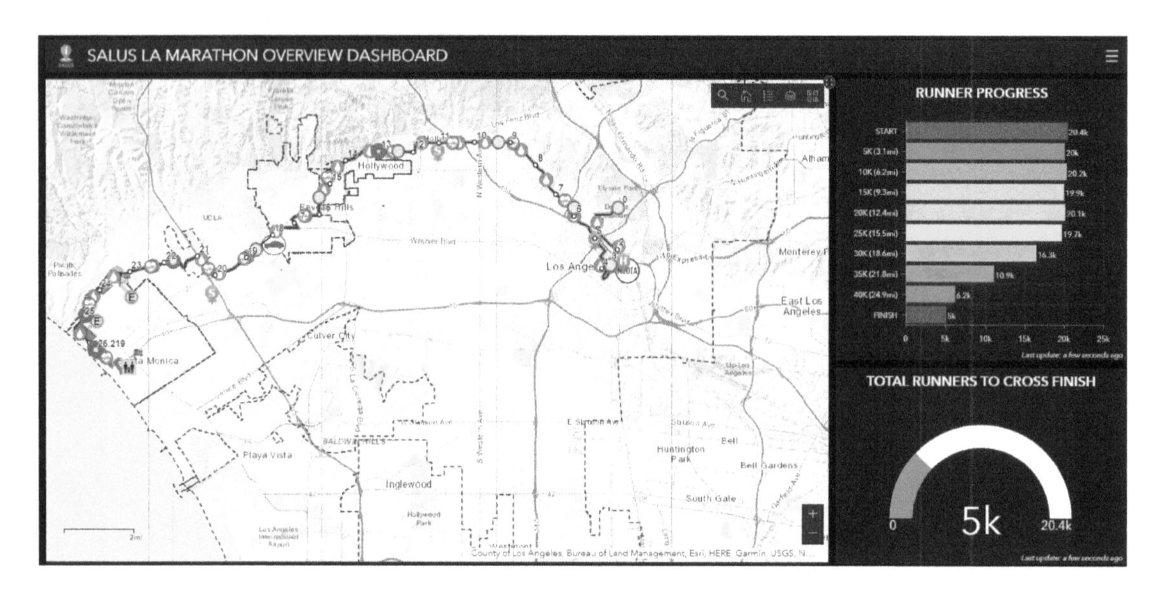

Figure 6.4. This dashboard from HSAC's mobile tracking and reporting app, SALUS, helped emergency responders and supporting agencies transform how they managed the 2019 Los Angeles Marathon. The app uses GIS to help identify threats, hazards, affected populations, real-time locations of staff, resources, and runners and has many other capabilities related to crisis management.

Adaptation: Responding to real-time changes

Each emergency is unique, and each presents unforeseen challenges. An earthquake might destroy a key bridge into a community. A fire may send burning trees onto a designated escape route. Flooding might slow or even prevent responders from reaching people trapped in their attics. Responders invariably encounter missing street signs, downed power lines, flooded streets, and other barriers as they struggle to navigate unfamiliar terrain. Response teams typically only become aware of these challenges once they encounter them in the field, leading to confusion, disrupted missions, and shifting priorities. This reality brings emergency responders to the adaptation stage of the collaboration process, when they add data about these unforeseen challenges to the information feedback loop established at the outset. This new information helps responders allocate resources efficiently and effectively.

Detailed, location-based scenario planning is critical; however, responders still must map and adapt to the changing conditions they encounter or failed to anticipate. American Red Cross and Federal Emergency Management Agency (FEMA) responders in Puerto Rico faced this obstacle during Hurricane Maria response efforts.[3] Volunteer cartographers across the world helped responders overcome the lack of spatial data by reviewing satellite imagery and mapping data prior to the hurricane landing and then comparing those images to data after the disaster to provide responders with new maps. Responders could see current data on their mobile devices and relay information about blocked roads or damaged bridges to the command center. This ad hoc, real-time assistance underscores the importance not only of volunteers in disaster response but also of data drills and good location intelligence as part of the planning process.

An emerging set of apps allows volunteers to share critical insights about what is happening on the ground. Emergency managers understand that not all the information coming from these apps is accurate and not all the apps are reliable in an emergency when communications can be disrupted. Nevertheless, apps such as those listed here can prove invaluable to cross-sector leaders.

- **For immediate assistance:** Crowdsource Rescue helps rescue efforts through the concept of neighbors rescuing neighbors. The app was developed as Hurricane Harvey hit Houston too quickly to connect professional first responders to everyone in need. The app uses GPS tracking and allows civilian volunteers, such as those who may have a boat, to sign in and help evacuate others.

- **For updates:** Weather Underground provides hyperlocal forecasts and updates on current conditions from local weather stations with customizable alerts. The app also allows users to report local weather changes and identify hazards to help inform others in the community. The map interface is interactive, allowing users to choose different layers to view elements such as rain accumulation, crowd reports, and satellite images. It also allows users to track storms in real time.
- **For communication:** Zello is a push-to-talk app, sort of like a walkie-talkie. Unlike walkie-talkies, there are no limits to the number of users or channels. It's available worldwide, wherever there's Wi-Fi or data service, and can be used like a two-way radio to communicate with family members or rescuers.
- **For updates and communication:** Hurricane, by the American Red Cross, allows users to monitor conditions in their area and let family and friends know they're okay with the use of a customizable I'm Safe alert for Facebook, Twitter, email, and text. The app includes a map that lets users track the hurricane's path and find Red Cross shelters in the area. In the absence of data connectivity, the app also provides step-by-step instructions of what to do before, during, and after a storm. It also includes a Hurricane Toolkit, with a strobe light, flashlight, and audible alerts.
- **For well-being:** The American Red Cross offers a variety of other apps that can be useful during hurricanes and other natural disasters. Its various apps educate people on how to prepare for floods, handle common first aid treatment (including pet emergencies), and respond to other emergencies.
- **For response:** Esri's Solution Templates for Emergency Response offer a variety of applications to help assess the full impact of an incident, return vital systems, execute service, and implement site restoration plans.

Local governments, agencies, nonprofits, and businesses also create apps to deal with specific situations. For example, during Hurricane Harvey in 2017, the City of Houston partnered with Harris County to release information on flooding estimates. Then, during the recovery phase, the city updated the status of city services, tracking power outages, transit routes resuming service, and more.

During Hurricane Irma in 2017, Miami curated flood maps from Miami-Dade County's website and initial storm surge information from Florida International University's *Storm Surge Simulator*. Information on city services, transit, and other critical developments throughout the response and recovery complemented the efforts of the flood mapping. Miami provided nine pairs of staff members from the Department of Planning and Zoning and four volunteer teams with ArcGIS Survey123®, a tool that allowed users to input flood data in the field from their smartphones. "Within 24 hours, we had more than 700 data points," said Michael Sarasti, chief innovation officer for the City of Miami.[4]

Cities can also expand the range of data they use to inform responses. One agency may have information that will accelerate response by others. For example, a state Department of Motor Vehicles can identify resources that might assist residents who require help finding transportation out of the city. These kinds of cross-agency predictive projects require a culture of data sharing in city government.

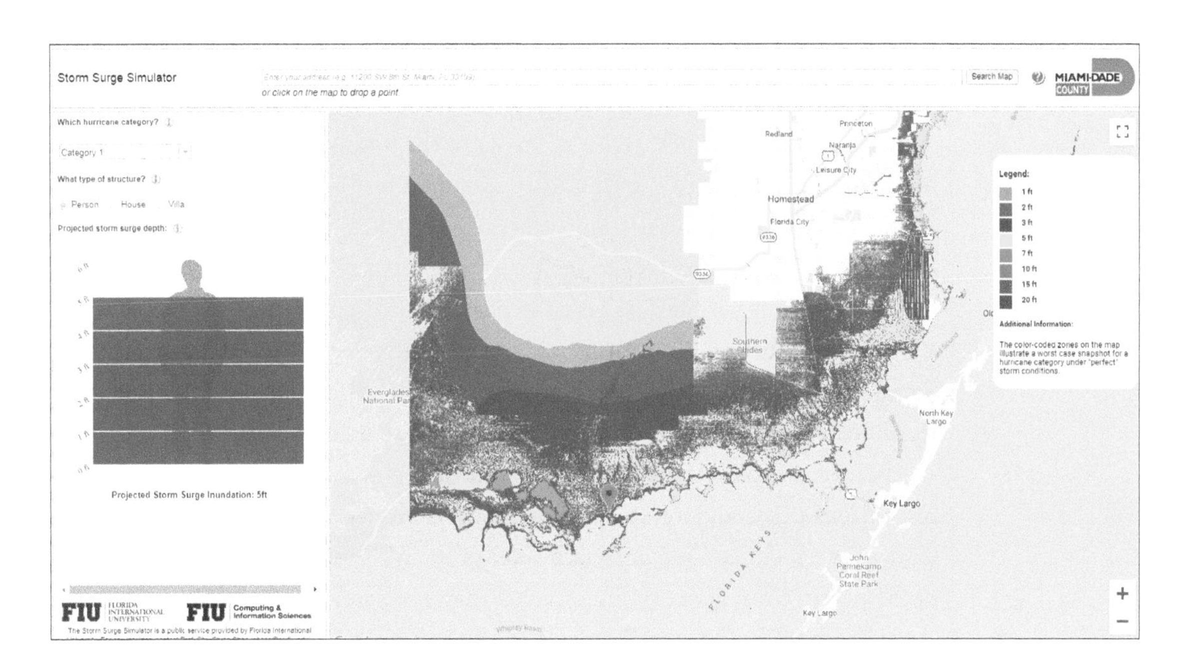

Figure 6.5. Florida International University's Storm Surge Simulator highlights storm surge levels in selected areas. The map also features a tool that helps users better understand risk by visualizing storm surge levels in relation to the average height of a person or house.

Mapping and data lead the way in the search for missing Flight MH370

This section will examine the use of GIS mapping tools during the search for a passenger airliner that disappeared—the tragic flight of MH370. At first glance, it may seem counterintuitive to review the circumstances of a massive search that did not find its target. But, as one of the most extensive geospatial search efforts ever launched, this multiyear, multinational collaboration offers an instructive example. The geographic, structural, and intangible lessons learned from this unique disaster response can help cities, states, and other responders as they prepare for unforeseen emergencies that are sure to happen.

On March 8, 2014, at 12:41 a.m. in Kuala Lumpur, Malaysian Airlines Flight MH370 took off en route to Beijing with 239 people aboard. Just 38 minutes later at 1:19 a.m., copilot Fariq Abdul Hamid sent a parting message to air traffic controllers as the plane exited Malaysian airspace: "All right, good night," he said. At 1:21 a.m., the plane's transponder, the mechanism that sends identifying information to air traffic controllers, shut off. Just a minute later, Thai military radar lost contact with the plane. At 1:28 a.m., radar in the Southern Surat Thani province of Thailand picked up an unidentified aircraft after the plane appeared to have altered its course entirely. Radar would lose contact again just minutes later and the plane would not get picked up again until 2:15 a.m. as it flew over Malaysia's Pulau Perak. This was the last time the plane would appear on anyone's radar.[5]

After exhausting all options, air traffic controllers notified Malaysian Airlines that MH370 had fallen off their radar.[6] Satellite data indicated that the plane remained in the air at least six hours after it lost radar contact.[7]

Before geospatial or communications data was fully incorporated into the response, surface and air search teams began immediately scanning the South China Sea, Andaman Sea, and both sides of the Malaysian Peninsula for any sign of the plane. Once government officials analyzed satellite communications data more thoroughly, the search shifted to the southern Indian Ocean about 2,500 kilometers off the coast of Perth, Australia. It was here that the Australian Maritime Safety Authority (AMSA) took the lead, coordinating search efforts. From March 18 to April 29, 2014, AMSA led a search spanning 4.7 million square kilometers of ocean. Over the course of 3,177 hours, AMSA coordinated 21 aircraft on 345 flights over the southern Indian Ocean. Additionally, 19 ships from eight countries searched the area.

Search and rescue workers depended on geospatial data throughout the formation, operations, and adaptation stages of this collaboration that continued for several years. Geospatial data became the basis for coordinating resources and maximizing utility of the tools and expertise of the critical stakeholders involved at all stages of the search.

Formation: Narrowing the search area and assembling the team

Coordinating these efforts took immense effort. Narrowing the search area for MH370 was a prerequisite to the possibility of finding the plane in a massive span of ocean, but also to optimizing the human capital and physical resources poured into the search.

To begin, search teams mapped the plane's probable flight paths based on an analysis of satellite communications, aircraft performance, and radar data collected by a team of international experts known as the Search Strategy Working Group.[8] They drew a line across the probable flight paths at the point where the plane likely would have run out of fuel and begun its descent. This line became known as the 7th arc. With an aircraft glide range of 30,000 feet, the analysis then determined that the plane was likely within the range of 31.6 miles west and 28.8 miles east of the 7th arc between latitude 20 and 40 degrees south—an area of about 232,000 square miles.[9] During this process, the Australian, Malaysian, and Chinese governments established the Joint Agency Coordination Centre (JACC)[10] to maintain communications and support the effort. With the search area determined, the JACC transitioned the search into its operation stage.[11]

Operations: A bathymetric survey of seafloor topography

With the search area set, AMSA took the lead, deploying the Australian Transportation Safety Bureau (ATSB) to conduct a bathymetric survey of the ocean floor. Bathymetry measures the depth of a water body and maps the underwater features.[12] Using sonar equipment to bounce signals off the seafloor, the survey provided search teams a full view of the seafloor topography and identified potential hazards that could threaten response teams.

The survey was the first step of a two-phase effort to map the seafloor. In what became the largest marine survey ever conducted, ATSB teams collected 107,000 square miles of bathymetric data, including data collected in the journey from port to the search area. Sonar equipment measured the strength of the return signal when a signal bounced off the seafloor, a metric known as backscatter data.

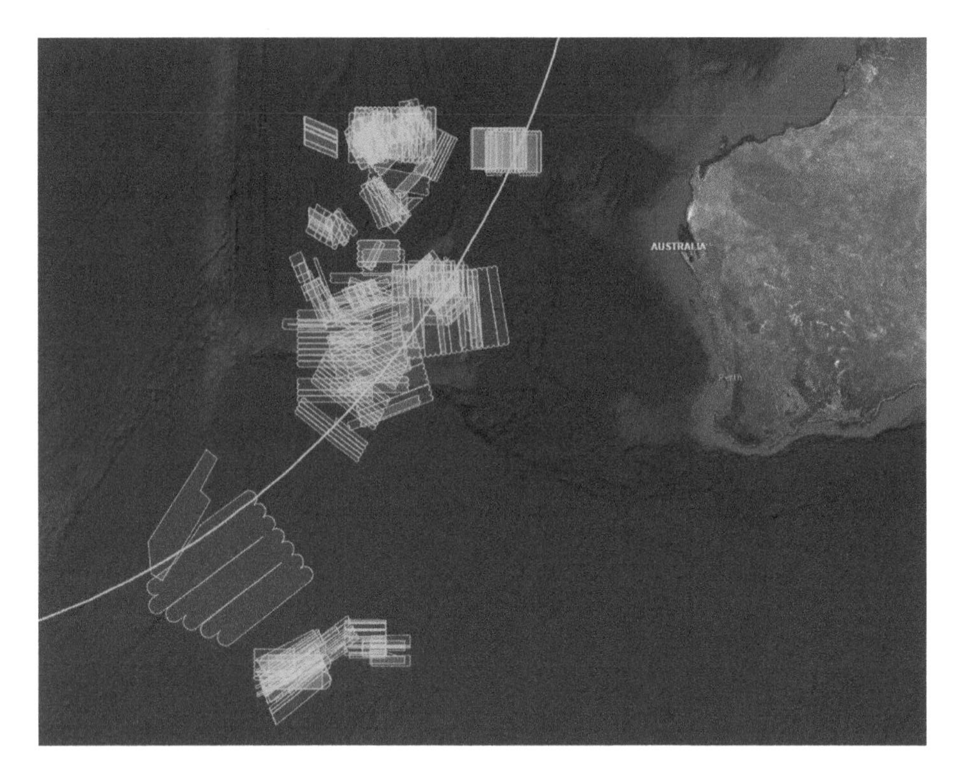

Figure 6.6. The initial surface and air search focused on the areas on either side of the Malaysian peninsula, in the South China Sea and the Andaman Sea. Satellite communications data analysis, however, led to the search being shifted to the southern Indian Ocean, where it was coordinated by the Australian Maritime Safety Authority. From March 18 to April 29, 2014, this search scanned vast areas of the ocean and involved 21 aircraft that conducted 345 individual flights over 3,177 hours, as well as 19 ships from eight nations.

These measurements helped determine how solid the seafloor below a search vessel was, which helped analysts distinguish between soft sediment and harder surface, such as rock. It also provided insights into certain seafloor anomalies.

The resulting map of the seafloor revealed deep canyons, underwater mountains (or seamounts) nearly a mile high, and landslides that stretch for miles. Compared to low-resolution satellite imagery previously available, this new map had 15 times higher resolution. It enabled underwater search teams to navigate the ocean safely as they moved on to the adaptation stage of the response.

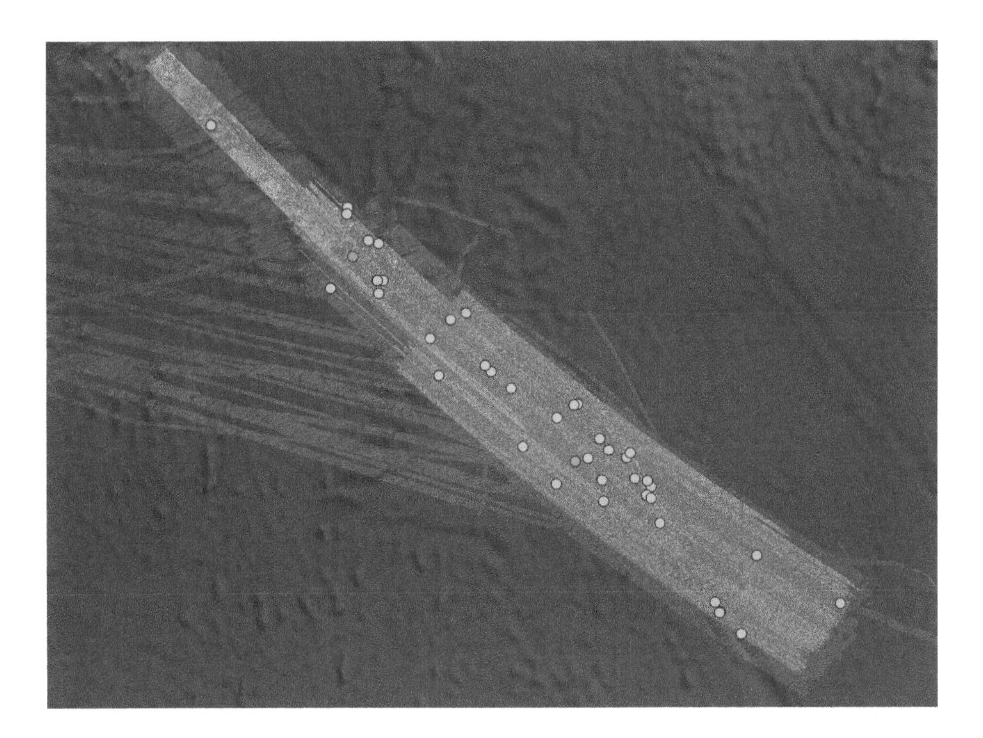

Figure 6.7. Because side-scan sonar data has a resolution of less than three feet, it can detect smaller features on the seabed and identify areas that differ from the surrounding environment. During the search, contacts identified by side-scan sonar were classified into different categories. The pink dots on this map represent contacts of high interest that warranted further investigation, whereas the green dots represent contacts of interest that were unlikely to be significant to the search.

Adaptation: Analyzing seafloor features for signs of MH370

Using the data from phase one of the search, underwater vehicles navigated unfamiliar terrain carrying side-scan and multibeam sonar equipment to get even higher resolution seafloor maps. Side-scan sonar measures the strength of the return signals of high-frequency sound pulses and has a resolution of less than one meter, helping scientists locate smaller items (such as luggage, airplane seats, and so on) and further differentiate between soft, hard, and irregular features. The ATSB classified the collected data in this phase of the search according to its probability of significance to the search.[13]

Of the more than 46,000 square miles of data collected in this phase, 626 data points were identified as of some interest but low probability of significance to the search, helping to eliminate a significant amount of effort and conserve resources in the underwater search.[14] In high interest areas, underwater vehicles dove deeper to investigate and officially identify the objects found. Sonar, photographs, and videos captured sunken ships, whale bones, and steel cable from vessels, among other items.[15]

Unfortunately, despite valiant efforts from everyone involved, searchers did not find MH370. On January 17, 2017, the Ministers of Transport from Malaysia, Australia, and China issued a joint statement suspending the search.[16] After two and a half years, the search for MH370 had cost roughly US $151 million.[17]

As of this writing, 32 items of debris from the flight have been recovered after washing up on the shores of South Africa, Madagascar, Mozambique, Pemba Island, Réunion Island, and Mauritius and brought to the attention of the MH370 search team. Of those items, 11 were not identifiable, three were confirmed, three were likely, eight were highly likely, and seven were almost certain to have come from MH370.[18] Australia's Commonwealth Scientific and Industrial Research Organization (CSIRO) used the locations of the recovered debris to conduct

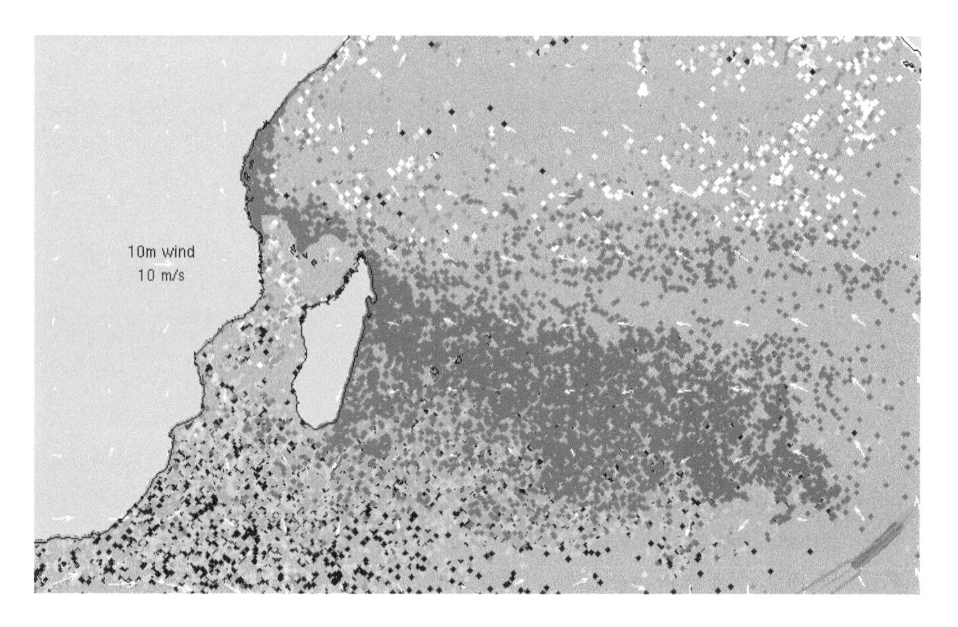

Figure 6.8. Researchers modeled potential locations of debris based on ocean currents. The colored dots on the map show the potential locations of debris if the plane went down at a location on the flight arc of the corresponding color.

a drift analysis in the hopes of revealing new information about the location of the plane.[19] The results of CSIRO's drift analysis were consistent with the initial determined search area, increasing confidence in the chosen location of the initial search efforts.[20] Using this data, a marine robotics company, Ocean Infinity, received a 90-day window starting in January 2018 to try to find the plane.[21]

The searchers needed to determine not just where the plane may have gone down but where the debris would be still floating at the time of a search or would have sunk and dispersed. Experts reconstructed the fight path and then modeled the drift patterns, but ultimately, searchers did not find the airliner.

Lessons for collaboration: Preparing for disaster

Cities and states rarely incur a large air disaster, but they can derive valuable lessons from the search for MH370. First, for AMSA, just like a local emergency management office, geography controlled many of the factors related to the formation of the collaboration, including who to invite, what resources they would contribute, and their geographical reach. Beyond simple proximity to the disaster, identifying key partners involves a complete understanding of the human and capital assets owned by the potential partners. In this case, the Australian government had the most resources closest to the search area and took the lead in coordinating public, private, and nongovernment resources according to the search's needs.

Second, at the onset of the emergency, Malaysian and AMSA officials began gathering all spatial data and relevant maps. As with the NYC data drills, this process shows leaders how the quality of the existing data and the speed with which it can be assembled has real consequences in cutting response times. Maps that defined the search area helped create a shared understanding of the scale and the assets required to address the disaster.

Third, the search illustrated the importance of location intelligence in the collaboration's operation. Responders needed to augment the maps, as would be the case in any domestic US disaster. Thus, the map must incorporate and visualize all critical data relevant to the preparation for the disaster and all incoming data after a disaster strikes. It must also be well-designed and actionable enough for users to quickly digest information they see on the maps and make important decisions. In the search for MH370, partners visualized and shared geospatial and bathymetric information in real time while using vast amounts of data to gather insights from experts across sectors. Using data from various sources on a cross-functional platform facilitated the work of hundreds of individuals, helping to guide action and enhance the effectiveness of the collaboration. Additionally, the classification of data touch points from the bathymetric survey helped with prioritization and search decisions.

Fourth, the quality of the map visualization—its transparency and accuracy—can create the intangible benefit of increased trust among critical stakeholders. When partners can easily contribute what they discover on the scene while simultaneously seeing the same from others, the transparency increases trust and confidence. This gives way to openness, understanding, and cooperation between response teams—the lack of which often hinders disaster response. Although the effort of recovering Flight MH370 ultimately proved disappointing, maps enabled the collaboration to better allocate and optimize resources, make strategic decisions, communicate, and adapt to change in real time. It is not impossible that, with an ongoing deployment of data, maps, and collaborative resources, and with changing oceanographic conditions, MH370 may yet be found.

Notes

1. Stephen Goldsmith, "Beyond Duck and Cover," *Government Technology*, March 2018, https://www.govtech.com/data/Beyond-Duck-and-Cover.html.

2. Amen Ra Masahki, personal interview, March 4, 2019.

3. Mimi Kirk, "When Cartography Meets Disaster Relief," *CityLab*, October 11, 2017, https://www.citylab.com/environment/2017/10/how-open-source-mapping-helps-hurricane-recovery/542565.

4. Chris Bousquet, "Data-Driven Emergency Response: Learning from Hurricanes Harvey and Irma," *Data-Smart City Solutions*, October 3, 2017, https://datasmart.ash.harvard.edu/news/article/data-driven-emergency-response-learning-from-hurricanes-harvey-and-irma-113.

5. "Timeline of MH370 disappearance," *CNN*, updated January 17, 2017, https://www.cnn.com/2017/01/17/world/asia/malaysia-airlines-flight-370-timeline/index.html.

6. Ibid.

7. Australian Government, "The data behind the search for MH370," https://geoscience-au.maps.arcgis.com/apps/Cascade/index.html?appid=038a72439bfa4d28b3dde81cc6ff3214.

8. Australian Government: Department of Infrastructure, Regional Development and Communications, "Joint Agency Coordination Centre," updated November 14, 2018, https://infrastructure.gov.au/aviation/joint-agency-coordination-centre.

9. Australian Government: Australian Transport Safety Bureau, "The Search: Maps," updated January 8, 2019, https://www.atsb.gov.au/mh370-pages/the-search/maps.

10. Australian Government: Department of Infrastructure, Regional Development and Cities, "Joint Agency Coordination Centre."

11. Australian Government, "The Data Behind the Search for MH370."

12. United States Geological Survey, "Bathymetric Surveys," *Ohio-Kentucky-Indiana Water Science Center*, https://www.usgs.gov/centers/oki-water/science/bathymetric-surveys?qt-science_center_objects=0#qt-science_center_objects.

13. Australian Government: Australian Transport Safety Bureau, "MH370: Sonar Contacts," updated October 19, 2016, https://www.atsb.gov.au/publications/2015/mh370-sonar-contacts.

14. Australian Government, "The Data Behind the Search for MH370."

15. Australian Government: Australian Transport Safety Bureau, "MH370: Sonar Contacts."

16. Dato' Sri Liow Tiong Lai, Darren Chester MP, and Li Xiaopeng, "MH370 Tripartite Joint Communiqué," Ministry of Transport, Malaysia, January 17, 2017, http://www.mot.gov.my/en/Kenyataan%20Media%20MH%20370/MH370%20Joint%20Communique.pdf.

17. Juliet Perry and others, "MH370: Search suspended but future hunt for missing plane not ruled out," *CNN*, updated January 17, 2017, https://www.cnn.com/2017/01/17/asia/mh370-search-suspended/index.html.

18. Malaysian ICAO Annex 13 Safety Investigation Team for MH370, "Summary of Possible MH370 Debris Recovered," Ministry of Transport, Malaysia, updated December 30, 2018, http://www.mot.gov.my/en/Laporan%20MH%20370/Summary%20of%20Debris%20Recovered%20-%20Updated%2030%20Dec%202018.pdf.

19. "Drift Modelling Simulations by CSIRO," YouTube, uploaded by Geoscience Australia, June 6, 2017, https://www.youtube.com/watch?v=BlGgudkQ7Dw#action=share.

20. Australian Government: Australian Transport Safety Bureau, "Assistance to Malaysian Ministry of Transport in support of missing Malaysia Airlines flight MH370 on 7 March 2014 UTC," released July 30, 2018, https://www.atsb.gov.au/publications/investigation_reports/2014/aair/ae-2014-054.

21. Ocean Infinity, "Ocean Infinity to Continue Search for Missing Malaysian Airlines Flight MH370," October 1, 2018, https://oceaninfinity.com/2018/01/ocean-infinity-to-continue-search-for-missing-malaysian-airlines-flight-mh370.

7

Increasing sustainability

The way natural systems work to stay in balance is one way to look at sustainability. Another way to look at sustainability is to consider how human society lives in relation to those natural systems. The World Summit on Social Development in 2005 identified the three pillars of sustainability as economic development, social development, and environmental protection.[1] These pillars aim to support our present-day needs without sacrificing the ability of future generations to live in harmony with the world around them.[2] This chapter looks at how collaborations address the issue of increasing sustainability in the context of climate change and other serious environmental challenges.

Cities working together

In 2006, the world's largest cities created their own collaboration by participating in the C40, a growing network of megacities addressing climate change. In 2007, New York City Mayor Michael Bloomberg released PlaNYC, a comprehensive sustainability plan to deal with climate at the local level.[3] That same year, the Harvard Kennedy School honored Seattle Mayor Greg Nickels for his city's environment agenda to address climate issues locally. The Seattle plan started to take shape when a drought caused the city to realize that reduced snowmelt jeopardized its reservoir, drinking water, and hydroelectric power, and increased the fire risk.[4]

C40's mission to reduce greenhouse gas emissions and other climate threats, and the work in Seattle and New York City, inspired other cities to become more active in structuring collaborations that could benefit the environment. The C40 Urban Climate Action Impacts Framework describes the vast stakeholder network and lists areas of benefit, including improved health outcomes, air quality, livability, and economic competitiveness.

Cities built the C40 framework to help them address the interconnectivity and complexity of different types of environmental impacts affecting air, water, noise, and biodiversity, while also providing a tool to measure progress. This framework can be accomplished through actions such as mapping needs, identifying critical stakeholders or resources, and working together on specific projects such as remediating lead in paint or soil, building a park, or planting trees.

CitySwitch Australia

Environmental collaborations come in different forms and scales, from national and multinational to hyperlocal and regional. One model consists of a partnership of multiple cities, with each city forming its own collaboration with local nonprofit and business partners. Funded by Sydney, Melbourne, Adelaide, Perth, and several local governments in Australia, the CitySwitch Green Office uses this model. The collaborative supports efficient energy use for commercial office tenants in Australia. CitySwitch establishes green infrastructure best practices and promotes energy efficiency initiatives in the private sector with the goal of protecting the environment while generating a positive return on investment.

One product of CitySwitch involves mapping specific building-related renewable energy opportunities. In figure 7.1, CitySwitch Melbourne identifies rooftops that could support green infrastructure or solar power.[5] This idea is similar to a New York City project in 2011 that used lidar to identify rooftops that would support solar. One can imagine a range of map overlays that depict renewable energy progress among buildings in a city or between multiple cities. These overlays could also identify additional opportunities, track retrofits and reduced energy use according to building size, and encourage friendly competition among building owners and cities.

Another Australian effort features a cross-sector collaboration to improve the structural integrity of buildings. The National Strategy for Disaster Resilience involves key players in the Australian insurance and construction industries that map building vulnerabilities and advocate for stronger building codes. This collaboration encourages action through offers of reduced insurance premiums.

Efforts to address the complexity of the environmental challenges must involve multiple entities acting together in their own organizational ecosystem. Formation of these partnerships, such as the ones involving CitySwitch and National Strategy for Disaster Resilience, often involve the same entities participating with each other in more than one effort.

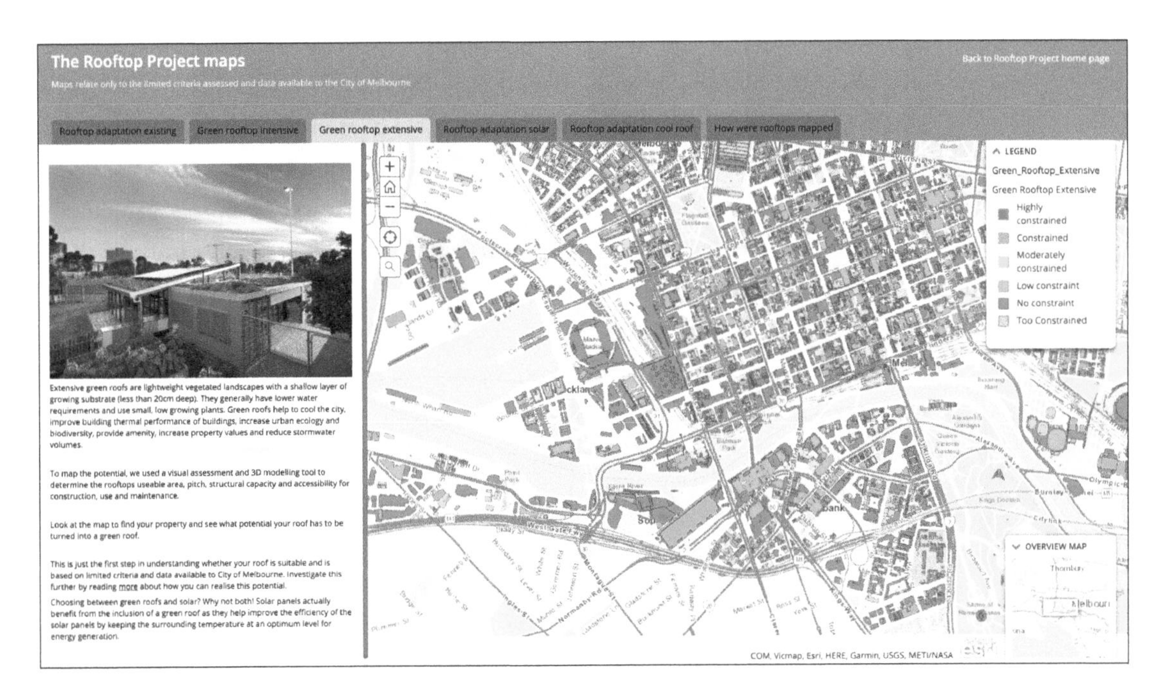

Figure 7.1. By mapping all the rooftops in the city of Melbourne to see if they have the potential to be turned into solar, cool, or green roofs, Rooftop Project helps residents and businesses reimagine how they could use their roofs to increase sustainability.

Formation: Collaboration across boundaries

Forming a collaboration to improve environmental outcomes, such as improved air and water quality, often involves interstate or cross-national organizations. However, to achieve success in these areas, collaborations must take successful remedial actions at the local level. The case studies in the next section will illustrate the role maps play in the formation of these collaborations, how they inform resource allocation, and how they enable continuous adaptation.

Protecting the Mississippi River watershed

Watershed management, with its sprawling ecology, demands a broad multiparty response.[6] Effective action begins by mapping the location of key stakeholders that affect the hydrology of the river. Some watersheds fall exclusively within a single local jurisdiction, such as a county with a contained lake; however, watersheds more often cover multiple jurisdictions and include creeks, streams, rivers, and groundwater that often flows into oceans and rivers. Mapping watersheds visualizes the challenges facing the ecosystem, identifies parties important to watershed health, offers context for operational priorities, and supports real-time efforts to mitigate pollution.

The Mississippi River watershed, stretching through 31 states, is the largest watershed in the United States. Dozens of organizations dedicate themselves to various environmental activities along the 2,340-mile length of the Mississippi. Adequately protecting watersheds of this size requires dividing them into progressively smaller, local sub-watersheds. For example, in 2020, the Mississippi River Cities and Towns Initiative included nearly 100 mayors who collaborated with important federal organizations to undertake activities related to transportation, farming, industrial use, and other river management issues. Federal organizations involved include the Tennessee Valley Authority, the US Army Corps of Engineers, the US Geological Survey (USGS), and the Department of Transportation, among others. The Mississippi River Cities and Towns Initiative includes two levels of collaborative activity. The mayors collaborate with their peers to address common problems. But with support from organizational and federal partners, the mayors also manage their own local cross-sector collaborations that focus on the sections of the river over which they have jurisdiction.[7]

The existence of so many organizations working along the Mississippi underscores the need to ensure that their members work in concert to achieve individual local goals and support the goals of others. GIS maps that visualize their actions provide transparency and help members avoid duplication of effort and unintended consequences. In this regard, GIS information helps members integrate solutions that affect one another. For example, agriculture runoff into one area of the river affects the river's health in other areas. By visualizing these kinds of processes, partners can also map, plan, and coordinate more effective conservation practices across interlocking cooperatives.

On a larger scale, the USGS demonstrates the value of maps by charting nutrient pollution throughout the Mississippi River Watershed. Reducing excess nitrogen and phosphorus runoff during storms is a priority in maintaining the health of rivers. Excess nitrogen and phosphorous stimulate algae growth that hurts plants and wildlife and causes other problems. USGS maps support local groups and other federal agencies such as the US Department of Agriculture's (USDA) Natural Resources Conservation Service (NRCS) in its efforts to investigate and manage nutrient pollution.[8]

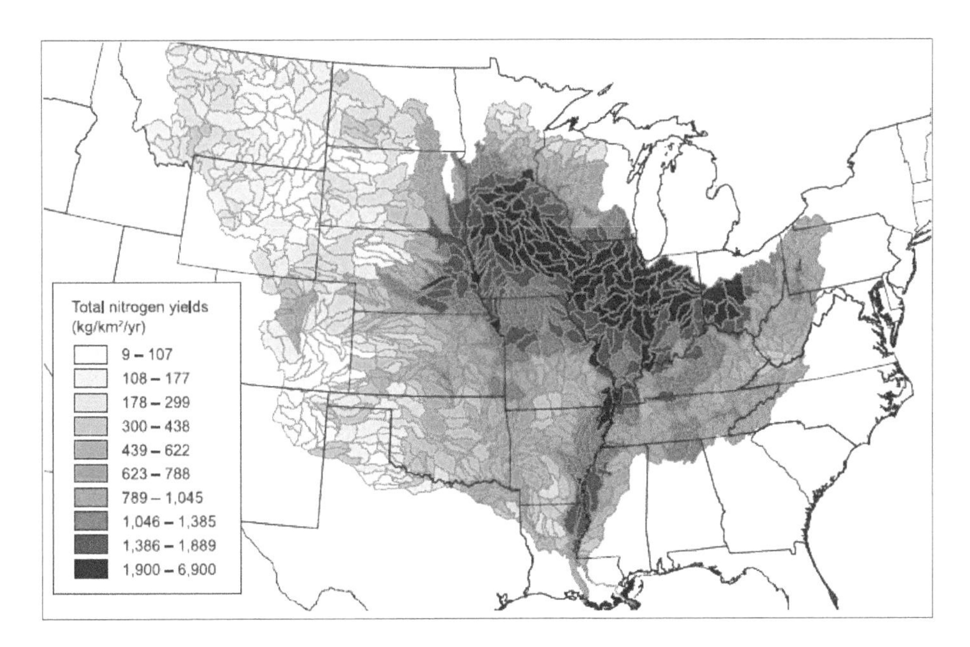

Figure 7.2. This map shows levels of nitrogen in the Mississippi River basin. Spatially referenced regression on watershed attributes modeling estimates that 60 percent of nitrogen in the Gulf of Mexico results from agricultural sources such as fertilizer. The USGS monitors phosphorus and nitrogen in the basin to understand rivers' roles and interactions with the nutrients.[9]

The NRCS provides watershed management best practice frameworks such as the Mississippi Watershed Adaptive Management model used by the Mississippi River Healthy Watersheds Initiative. This collaboration of nutrient producers and landowners concentrates on implementing voluntary conservation practices "that reduce nutrient loading, improve water quality, restore wetlands, enhance wildlife habitat, and sustain agricultural profitability in the Mississippi River Basin," according to an ArcGIS StoryMaps story by Geoplatform.gov. USGS efforts to continuously monitor and map the Upper Mississippi River basin illustrate the critically important role real-time data plays in the formation and operation stages of collaborations across a range of stakeholders including farmers and federal, state, and local agencies alike. For 30 years, the USGS has collected and mapped data on water quality, vegetation, and fish to support and improve local water quality standards. This helps various parties redirect their efforts in response to changes in the river that result from wildlife and human activity.

This approach to watershed management demonstrates the effective use of spatial analytics among multiple governmental levels. Data that facilitates the formation of a cross-sector collaboration can originate from a third party as a shared service supporting multiple organizations. As this chapter shows, the effective use of federal maps depends on how local collaboratives analyze and use the data to drive their actions.

Governments working together: Resilient Delaware

Sometimes, new laws call for the formation of a collaboration or cross-sector collaboration. For example, government can require a collaboration to protect a watershed from threats such as agricultural pollution, development, and storm water runoff. In such cases, volunteer and public organizations often join to take action such as cleaning debris from river basins or planting trees. These activities often support broader floodplain management goals that involve recruiting cross-sector partners to share resources, increase their collective capabilities, and enact protections beyond regulatory requirements.

The Delaware River Basin Commission (DRBC), a multistate and federal collaborative, is one such formally authorized agency that develops its own rules and governance structures beyond regulatory requirements. The commission includes more than 50 state and interstate agencies along the river. The commission's boundaries cut through several state and local jurisdictions along a 330-mile stretch from the river's headwaters near Hancock, New York, to the mouth of the

Delaware Bay. The DRBC, with its own authorizing statute, operates in a geography governed by other local, state, and federal environmental laws. For example, the Federal Emergency Management Agency (FEMA) controls flood insurance regulations that indirectly affect where a home can be built under certain structural and environmental restrictions.

To maintain a healthy river basin, the intertwined actions that affect its ecology and the sheer number of organizations involved require accurate and real-time spatial analytics. The ability to quickly locate illegal overflows, dumping, and other harmful actions supports mitigation and enforcement actions. For all these reasons, locational intelligence is critical in the planning and operation of a formal collaboration.

The DRBC supports many collective actions within its geographical boundaries. Consider the cities affected by the DRBC, which decided to require additional building height for structures inside the flood insurance mapping area. Planners needed accurate flood and engineering information to make complex decisions on imposing more costly building requirements. The interconnectedness of these decisions and consequences illustrates the need for cooperation. Decision-making in one area of the river affects another area. For example, a home that washes away in a flood can easily cause downstream damage. These cascading effects on an ecosystem show the importance of geospatial insights in determining the makeup of a collaboration and how its decisions affect property values, building safety, and so on.

Because the Delaware River's water levels and health change constantly, the DRBC's work in flood protection and watershed management requires a detailed and current understanding of locational risks. Monitoring and mapping water levels, river obstructions, and changes in the adjoining basin in near real time supports the commission's initiatives concerning water quality protection, water supply allocation, regulatory review (permitting), water conservation initiatives, watershed planning, drought management, flood loss reduction, and recreation.[10] Thus, the DRBC manages a network of organizations and volunteers to ensure broad participation and vigilance across impacted or at-risk areas. For example, nonprofit volunteers and government workers can upload time- and place-stamped photographs via GIS platforms, and field workers can enhance professionally drawn maps with real-time observations that support flood control and water protection efforts.

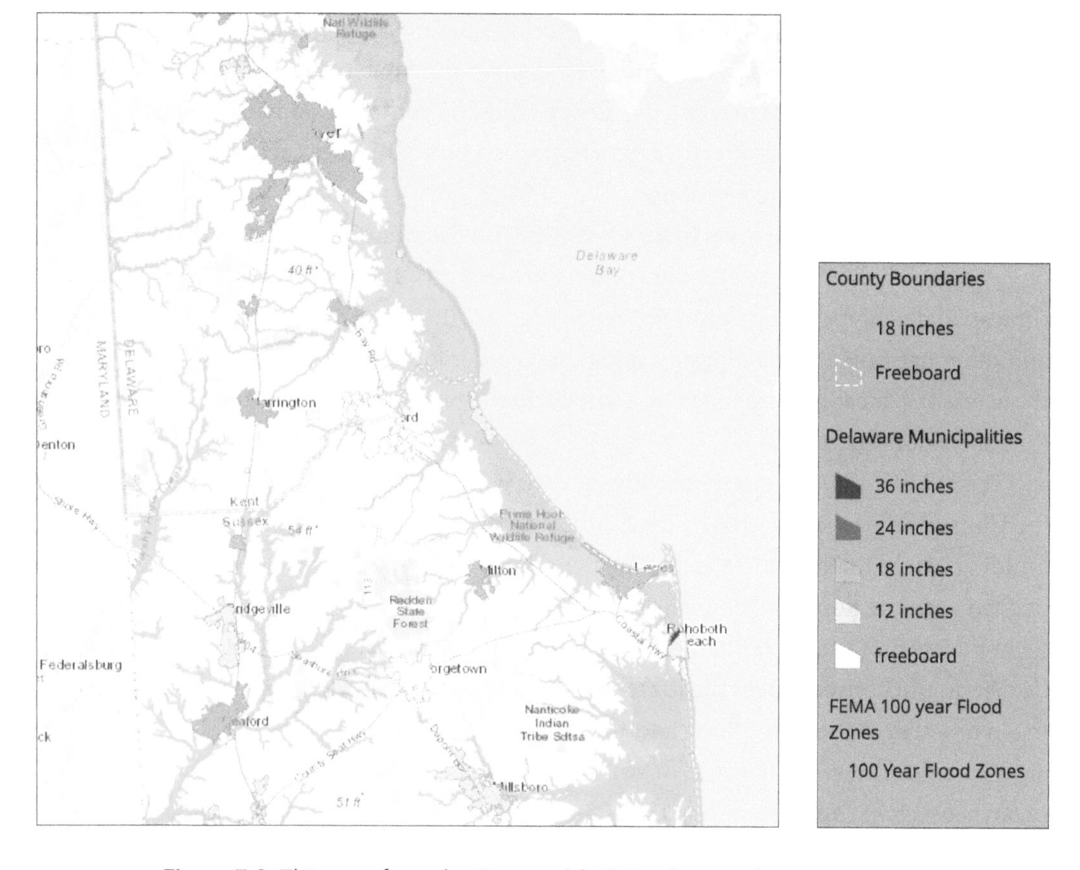

Figure 7.3. This map from the Sustainable & Resilient Delaware initiative demonstrates the need for implementing freeboard requirements in flood-prone communities by highlighting flood levels across the state. Freeboard is a floodplain management strategy requiring the bottom floor of buildings be elevated one to three feet above FEMA's National Flood Insurance Program (NFIP) minimum requirements. Elevating buildings above these minimums allows cities to account for the unknown (for example, changing floodplains) and increases resiliency against flooding.

Experts recognize that well-visualized mapping is essential for watershed and flood protection and for emergency response. However, they worry about how missing data can impact infrastructure investments and jeopardize mitigation efforts. Errors in 100-year flood designations can hurt flood protection efforts; underfunding can result in mistakes during mapping of changing environmental conditions. A recent Rand Corp. report underscored the role of mapping in watershed preservation efforts. Some data lacks information about existing countermeasures such as flood walls and reinforced structures to guard against flooding, high winds, and earthquakes. Learning about these countermeasures is important for planning as federal, state, and local governments and the private sector seek to improve resilience against natural disasters.[11]

Similarly, Colin Wellenkamp, executive director of the Mississippi River Cities and Towns Initiative, worries about the long-term effectiveness of the collaboration. Underfunding and strict regulations for federal flood risk mapping programs can affect map quality. Flood-mapping programs used by the Army Corps of Engineers, USGS, and FEMA must be augmented with real-time information on the ground, Wellenkamp said. The Cities and Towns initiative requires layered mapping of flooding, weather patterns, local conditions, and the location of debris that affects water flow and drains.

Generating more and better information about real-time and predicted conditions, through augmented maps and shared analytic insights, will improve the governance and effectiveness of the collaboration.

A partnership designed to induce community participation: Green your routine

Since 2009, the City of Fort Lauderdale has endorsed a range of sustainability activities promoting cross-agency action, citizen advisory boards, and sustainability planning. While not unique, this interagency collaboration shows how a group of volunteers works alongside city officials to engage the community in sustainability efforts.

As mentioned in chapter 2, Fort Lauderdale, a longtime leader in green initiatives, encouraged local involvement in environmental protection through the Green Your Routine website. This social platform allows residents, businesses, and government to educate each other on practical ways to protect the environment. Fort Lauderdale's volunteers identified activities in the areas of land use, waste reduction, transportation, and energy that enabled an individual, family, or business to participate in sustainable practices. The Green Your Routine interactive map tool displays 15 discrete environmental practices that city residents can add to their daily routines, including composting, building retrofits, and home energy improvements.

In addition to helping residents embrace sustainable practices, the Green Your Routine initiative evolved to add more partners and assume more responsibilities. The GIS platform identifies locations where citizens can plant trees to

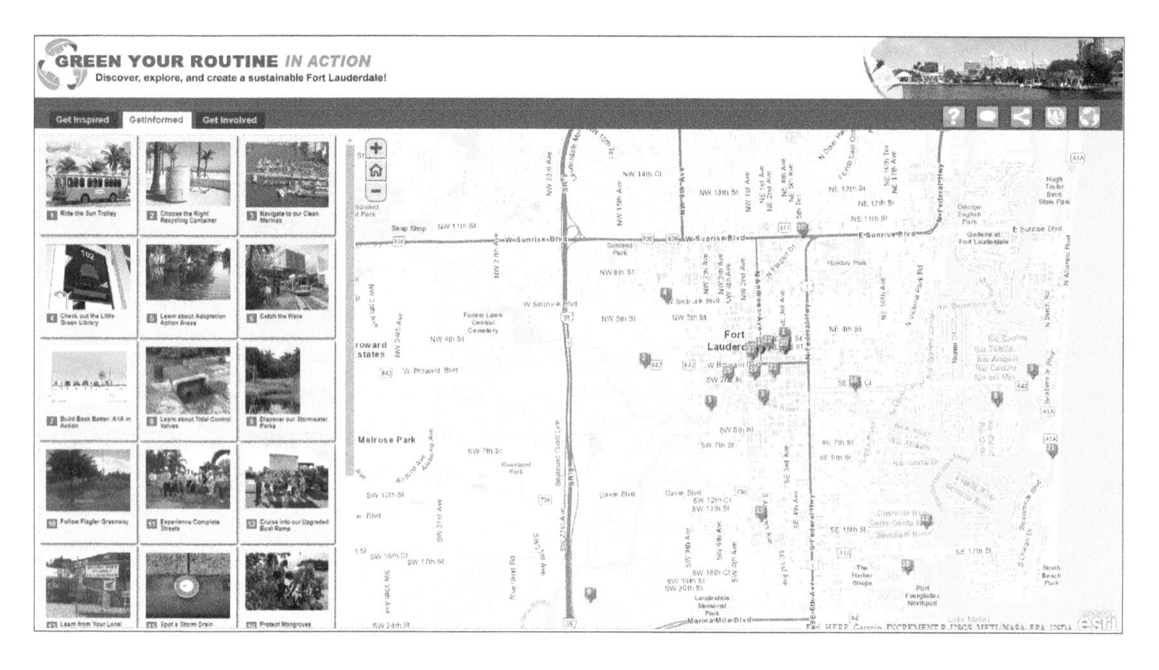

Figure 7.4. On the Get Informed tab of the Green Your Routine interactive map, residents can scroll over green markers that correlate to the menu on the left encouraging people to protect the mangroves, ride public transit, find the Little Green Library, and engage in other green practices.

reduce carbon emissions, improve air quality, and curb the heat island effect. Volunteers also helped identify locations for electric vehicle charging stations.

Ian Wint, GIS manager for the City of Fort Lauderdale, said the Green Your Routine map became an extension of the website, broadening the number of people who can participate. The picture-driven map identifies the relevant locations and what people should see when they arrive, Wint said. Engaging with the website often inspires people to get more involved in making their homes and offices more sustainable.

Fort Lauderdale increased its outreach by designing an interactive and well-visualized map filled with easily understood educational content that encourages practical sustainability practices. High-quality visualizations combined with actionable steps improved volunteerism while serving as a platform for individuals to collaborate with government, nonprofit volunteer groups, and neighbors. Green Your Routine shows how mapping visualizations help social platforms become important cross-sector assets. Ultimately, Fort Lauderdale's initiative lost momentum when the city did not approve more funding after initial grant funds ran out. This case demonstrated that mapping and an ad hoc set of relationships can induce a collaboration, but its growth and continuity require ongoing operational support from an entity at its center.

Organizing clean water volunteers: San Francisco's Adopt a Drain Program

In recent years, there have been significant breakthroughs in the use of platforms to deliver a product, create a social network, and bring people together around a goal. Breakthroughs that brought us Uber and Airbnb have also inspired a new form of cross-sector collaboration in which nonprofit organizations, government agencies, and volunteers join to share resources and solve public challenges.

The public can use relatively new applications, for example, to care for, improve, and provide feedback on the management of city infrastructure and natural assets, such as catch basins, parks, fire hydrants, and storm drains. In one example, the City of San Francisco took advantage of an app platform, spatial analytics, and volunteerism to manage urban storm water. In San Francisco, heavy rains can produce flooding as water rolls down large hills, carrying debris that ultimately clogs many of the city's 25,000 storm drains, creating potentially hazardous conditions. Jason Lally, open data manager at DataSF, worked with the city, its Public Utilities Commission, and Code for America to create Adopt a Drain, an app that enables residents to find their closest storm drain, claim it, and agree to clear debris in advance of storms. The city provides these volunteers with tools, supplies, and training.[12] As the volunteers became more involved with drainage activities, the city's Public Utilities Commission expanded the program. For example, volunteers adopted rain gardens, which are green areas built to absorb water from roofs, driveways, or streets to prevent runoff from reaching public drains. Participation in maintenance programs allows the public to monitor, report, and care for assets throughout their community. The city can use feedback from the app to improve daily operations and prepare for and respond to emergencies.

Cities also can use a generic app template to encourage volunteerism and civic engagement. As shown in figure 7.5, the Adopta app uses ArcGIS software to facilitate a comprehensive approach that allows any city to customize its asset adoption program for needs that range from unclogging storm drains to finding hydrants in a snowstorm.

Mapping brings parties together to achieve environmental goals, but the successful operation of a complex collective action requires accurate and reliable spatial analytics.

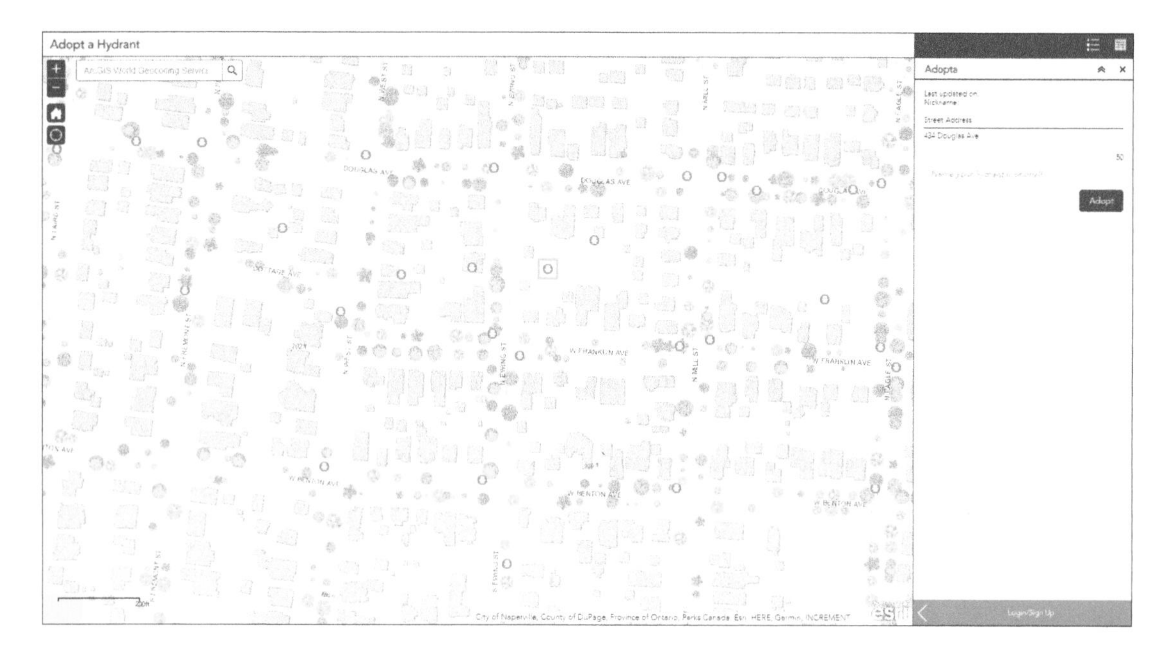

Figure 7.5. The Adopta configurable app can be used by government agencies and other organizations to engage the public in the operations and maintenance of natural and human-made assets (storm drains, hydrants, catch basins, parks, streams, trails, roads, vacant lots, and so on).

Operations: Bringing cross-sector actions together

The natural environment is a complex arrangement of interdependent ecosystems where events impacting one ecosystem can impact many. This interdependence means that addressing and preventing environmental hazards requires broad multiparty, interconnected, place-based approaches. This section discusses two examples of maps inspiring cross-sector action to address an environmental problem.

Cooling Dallas: A green and equitable response

One area of growing concern for public health and sustainability efforts in Sunbelt cities is localized heat islands. Large cities are landscapes of asphalt, containing tall concrete and steel buildings that create intense heat on the ground. Traffic congestion and resulting carbon dioxide emissions also trap heat,[13] creating a public health hazard. The negative effect of all these conditions tends to disproportionately impact poorer communities. Research has shown that green space and tree canopies help reduce temperatures in urban environments. Maps can

Figure 7.6. This map of Dallas identifies areas of high (orange) and very high (red) priority to reduce the urban heat island effect. Analysts used air and surface temperatures and data on poverty and other categories to prioritize populations most at risk from the harmful effects of heat islands. The city proposed increasing its tree canopy to mitigate the impact of heat waves and reduce summer energy use.

play a key role in the implementation of these methods by identifying areas in greatest need of green intervention.

The Trust for Public Land (TPL) and the Texas Trees Foundation (TTF) joined with city officials and other organizations in the Smart Growth for Dallas initiative, a program that addresses various issues of social, economic, and environmental resilience. In 2015, TTF laid the groundwork for the project to address Dallas' heat islands when it mapped the tree cover throughout the city. TTF used aerial imagery, and its volunteers physically identified and counted tree species in a sample of more than 600 plots.

Together, the groups examined the relationships among high temperatures, socioeconomic status, and health. Throughout the process, they relied on an intensive mapping exercise that overlaid locations of trees, income levels, public health information, temperatures in areas with high foot traffic, and levels of vulnerability to floods. The maps included streetscapes that examined where new pedestrian and bike lanes could be added safely and in a way that improves water management, while mitigating the dangers from heat islands. Volunteers also collected data showing areas of elderly concentration and schools without shaded playgrounds.[14]

Researchers layered a map of Dallas's parks with socioeconomic and demographic data from the US Census to better understand how the unequal availability of parks and trees affects low-income communities.

Robert Kent of the North Texas Division of TPL pointed to the importance of location intelligence in identifying the interconnected challenges of warming in an urban area. He noted that urban heat exacerbates cardiovascular conditions, asthma, and other health issues. In this context, the maps identified neighborhoods that suffer from the highest health disparities. More than just increasing tree canopy, the strategy calls for improved carbon-free transportation, better water management to absorb rainfall, and locating green infrastructure where it can best serve disadvantaged populations.[15]

The mapping exercise identified the Oak Cliff neighborhood as one of the highest priorities in terms of heat and health disparities. Oak Cliff lies southwest of downtown Dallas and has a population of 275,000, of which 28 percent live below the poverty line. Volunteers from Cool & Connected Oak Cliff—the civic engagement program led by The Nature Conservancy (TNC), TTF, and TPL—planted 1,000 trees to help reduce the heat island effect. The trees will remove almost 250 tons of carbon dioxide from the air and absorb 4 million gallons of storm water during the next four decades.[16]

Oak Cliff's 1,000 new trees will be just the beginning of the effort centered on that neighborhood and other parts of Dallas. TPL plans to update the map every year and monitor temperatures and public health data over the next five years. Continuous monitoring, use of GIS technology, and map visualizations will produce the accountability necessary to ensure that these efforts and their impacts continue. Beyond mapping disproportionate impacts on poorer communities, the collaboration also organized its operations around maps that highlighted progress such as where a tree canopy would make the greatest health impact. The maps also helped those managing the effort to identify groups and volunteers whose activities assisted in accomplishing the goals of the collaborative.[17]

Town-gown collaborations: A resilience toolkit for Richmond

The City of Richmond, Virginia, also used mapping and analytics to address the urban heat island effect. To identify which areas incur excessive heat, researchers with the Science Museum of Virginia in Richmond and Portland State University recruited volunteers to map local temperatures block by block. Volunteers included students from the University of Richmond, Virginia Commonwealth University, the Virginia Academy of Science, the Richmond Sustainability Office, and Groundwork RVA, a Richmond-based nonprofit that helps young adults respond to environmental and social issues. They also examined the relationship between temperatures, extent of impermeable surfaces, number of trees, and poverty rates. City officials focused on the apparent relationship between emergency medical responses and the hottest and most vulnerable areas.

By including surface temperature, tree canopy, concentration of impervious surfaces, and census data on income levels, researchers defined a quantitative index that helped identify the highest-risk neighborhoods. Areas with a below-average index score showed where the city could take actions such as planting more

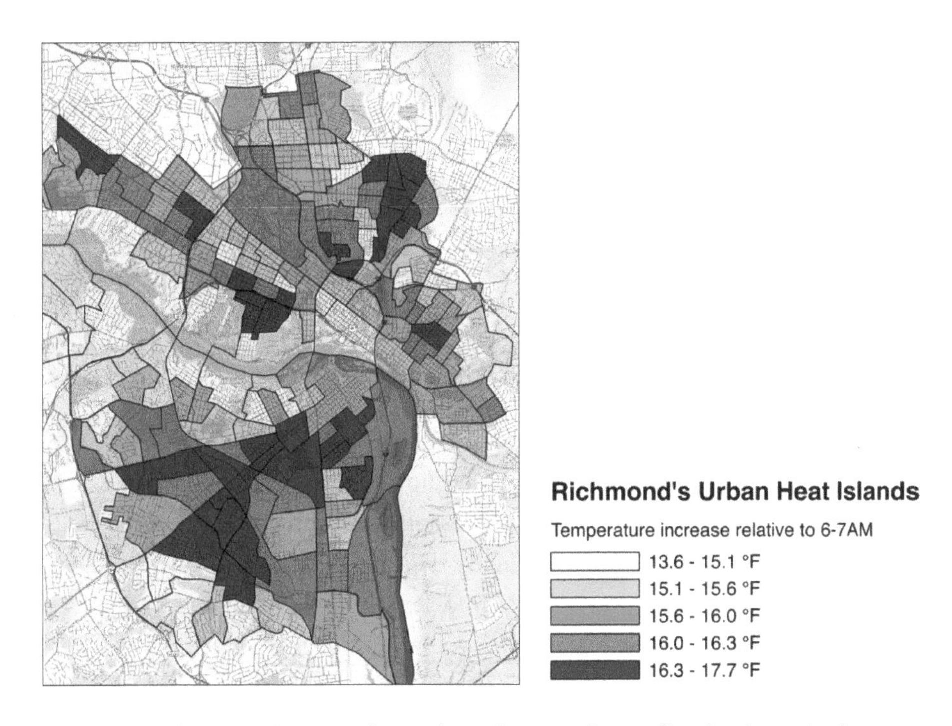

Figure 7.7. This map shows Richmond residents and city officials where the heat effects are worst. Inputs, arranged by census block, include data on impervious surfaces, income, and other risk factors.

trees, building structures that provide shade, and upgrading building construction requirements. These actions helped shape the US Climate Resilience Toolkit. They are part of an interagency initiative that operates under the auspices of the US Global Change Research Program managed by the National Centers for Environmental Information. The project incorporates a comprehensive group of partners who augment satellite images with mapping by others, including passengers in cars and bikes, who add information to the maps.

The next stage of this project—Steps to Resilience—examines mitigation scenarios for providing cool locations in the hottest parts of the city. Jeremy Hoffman, a scientist with the Science Museum of Virginia in Richmond, emphasizes the collective impact of the project when he describes the various connected actions of the partners, including installing low-cost air quality sensors for the hot spots, investigation of poor health outcomes, additional community garden and ground shade projects, and support of the city's sustainability initiatives.[18]

Storm water effects on watersheds: The Regional Stormwater Partnership of the Carolinas

City officials face few, if any, environmental challenges more expensive and serious than storm water and combined sewer overflows. The conversion of green space into impervious surfaces (streets and parking lots) pushes storm water directly into rivers, streams, or sanitary sewers. In almost 1,000 US cities, combined sewer systems collect rainwater runoff, domestic sewage, and industrial wastewater into one combined system. In large rain events, those sewers often overflow into streams and rivers because they cannot efficiently carry the volume of water to treatment plants, or the plants cannot process the volume. Storm water management problems have worsened when cities that have increasing storm intensity continue to permit paving and building over green space.

Storm water polluted with bacteria, antibiotics, nutrients, heavy metals, and other chemicals also presents potentially serious environmental and health risks, even when it does not overflow. Reducing the runoff produces healthier rivers and streams and better water quality.[19] However, finding efficient ways to reduce and channel the runoff presents challenges because of the wide range of stakeholders involved in causing and solving the problem.

With the support of the Environmental Protection Agency (EPA), cities increasingly look for ways to absorb water into the ground through green areas, holding ponds, porous surfaces, and more. These efforts involve nonprofits, cities, and developers who design new buildings; they also encourage businesses to build parking lots that reduce runoff and homeowners to act on downspout drainage, yard design, and upkeep. These actions depend on accurate mapping, topography, and hydrology to help drive the formation of the collaboration by identifying those who should participate. In this case, spatial analytics help the collaborative partners identify, analyze, and address the causes of storm water runoff.

The Regional Stormwater Partnership of the Carolinas (RSPC) supports a community of 18 municipal partners across North Carolina and South Carolina who joined together to implement storm water management policies. RSPC helps use resources for common activities among its members, for example, supporting mapping, training, and public awareness campaigns. The partnership facilitates real-time monitoring from field workers who use ArcGIS Online and mobile apps to input data and observations into maps as they walk along streams and evaluate erosion. Aided by spatial analytics, RSPC members act collectively to support these and other conservation efforts:

- Set storm water runoff fees based on the size and location of impervious surfaces.
- Determine the frequency and routes for street sweeping.
- Create capital improvement plans based on where investments produce the greatest impact.
- Determine the assignments of work crews charged with the construction and cleaning of swales and ditches.

RSPC is a large collaboration, but because of the depth, breadth, and complexity of storm water runoff problems, its city members collaborate on area-wide initiatives and regulations. They also collaborate to manage actions locally at the points where storm water becomes contaminated or joins the consolidated sewers.

Successful interventions require buy-in from residents and merchants. The list of action items proposed by RSPC to property owners illustrates the reach of its efforts:

- **Prevent muddy streets:** Construction sites can be managed to reduce muddy streets and, in turn, reduce clogged storm drains and other issues.

- **"Green" yard care:** Farmers and homeowners use fertilizers and chemicals that create pollution. County Cooperative Extension Services assists with test kits and advice to homeowners on best practices.
- **Recycle unwanted hazardous chemicals:** A single gallon of household hazardous waste can pollute more than a million gallons of water.
- **Reduce the volume of runoff:** Use water-absorbent building materials, build rain gardens, and plant flowers in depressed areas to absorb rainwater and prevent runoff.[20]

The RSPC collaboration shows that sustainability practices, even when concentrated in a particular community, affect many areas and involve many actors. Successful collective action uses location intelligence and timely, well-visualized information to identify businesses and nonprofits in specific communities that can help address local problems. They also use these tools to motivate volunteers,

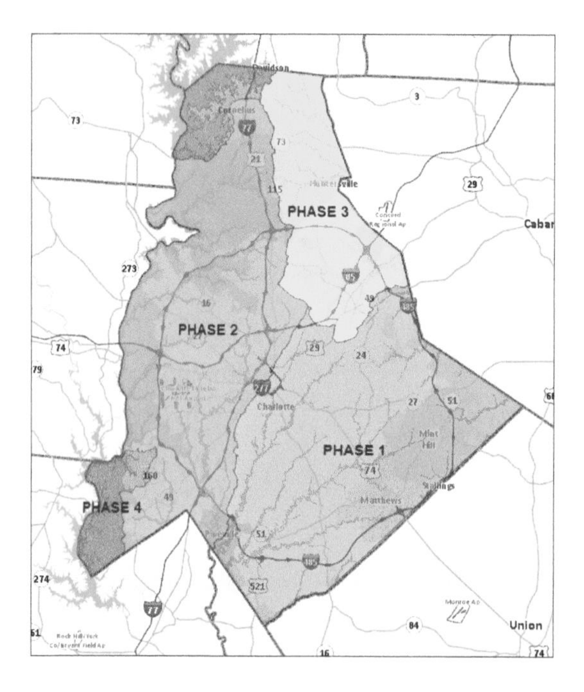

Figure 7.8. Greater Charlotte is remapping floodplains by zone. Remapping the floodplain is part of a larger effort in Mecklenburg County to improve the accuracy of its maps and help residents see how the changes in the map affect their properties and submit comments online.

locate areas impacted by heavy water runoff, and identify opportunities to add green spaces that will absorb the most surface water.

Cities across the country, with EPA encouragement, have implemented storm water fees that charge property owners based on the extent of their impervious surfaces, aiming to reduce water runoff. The entire framework of a storm water utility relies on mapping, including the way it sets up its fees, allows offsets, and fines polluters. Geographically oriented financial incentives motivate owners to better address storm water flows and consider activities such as increased tree plantings, swales, or new tile and concrete designs. One can imagine a public works director using information from flow meters, outflow events, runoff features, and the amount of hard surface area to better understand waste and storm water options. The better the mapping and clearer the visualization, the more effective the collaborative in addressing storm water runoff.

Adaptation: Working together to manage change

To be successful, cross-sector collaborations created to address environmental challenges must adapt to ever-changing conditions. Environmental changes affect who participates, what they do, and where they do it. This section discusses how public officials, scientists, and environmental groups use dynamic mapping to adapt to changing conditions that affect watersheds, including weather patterns, conservation practices, and development decisions. The NRCS in the Mississippi River watershed, discussed earlier, provides a good example of how adaptation works.

In the Mississippi River basin, financial and technical assistance is offered to farmers to address local water quality concerns that in turn further the larger goals in the basin, according to Dee Carlson, a water quality specialist with NRCS. NRCS and local affiliates work in this way because river ecosystems, which include adjacent floodplains, are complex and dynamic. Changes in this system can disrupt the connection between the river and its floodplain and adversely affect flood control, nutrient retention, water quality improvement, and fish and wildlife habitats.

Continuous mapping and monitoring of ecosystem conditions produce an ongoing inventory of nutrient data that, when visualized, helps partners better understand the impacts of various strategies on the watershed. This allows partners to adapt their operations to meet changing needs. The USGS Spatially

Referenced Regression on Watershed Attributes (SPARROW) models use technically advanced mapping such as watershed data and simple geological aspects to estimate the rate of absorption of certain contaminants, which researchers then map to alert the need for preemptive action. SPARROW mapping of hazards informs local leaders, who can then modify the steps they take to address water quality issues. Too often, cross-sector efforts operationalize a well-designed initial approach that does not consider ongoing change. Yet, whether the focus is on the environment, poverty, health, or some other condition, complex systems constantly change. The most successful multiparty efforts find a way to capture and share these changes in a way that allows constant adaptation.

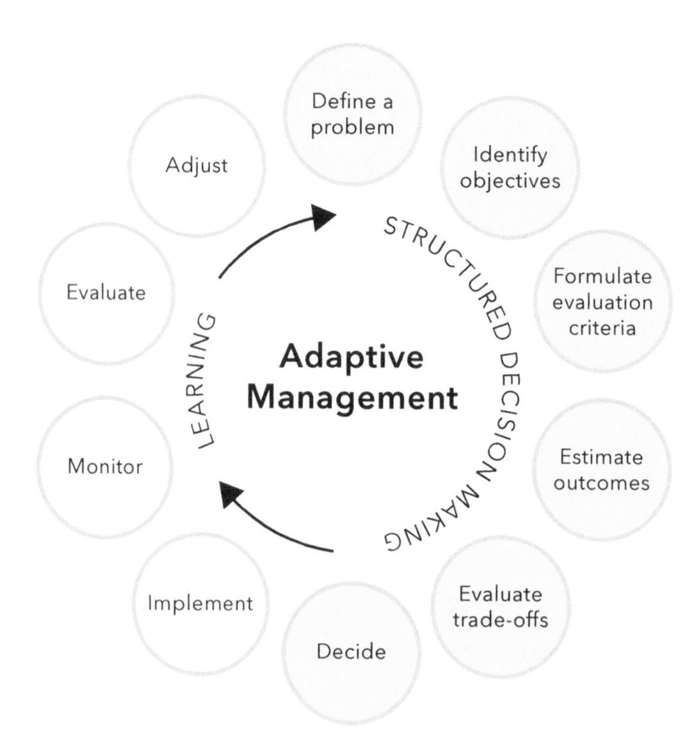

Figure 7.9. The USGS uses this adaptive management decision-making model to monitor and assess nutrient reduction practices and make iterative adjustments. This model produces a cycle of interventions that leads to new learning and opportunities.[21]

Lessons for collaboration: Mapping as a platform

The case studies in this chapter describe important actions taken by cross-sector partners to address the challenges involved in increasing sustainability. The initiatives combined public, private, and nonprofit sectors at multiple stages and produced results with several advantages:

- Improved spatial insights by combining national and local maps with real-time survey data and sharing it among the operating entities
- Assigned better-targeted actions to the collaborative and to individual partners
- Enhanced visualizations of actions and progress in real time to keep each party informed about each other's work, in a way that builds trust and enthusiasm
- Produced broader-based consumer actions through a catalytic narrative, for example, adopting a drain or not paving a driveway with impervious asphalt
- Created more accurate and dynamic tracking and monitoring of conditions based on observations recorded through mobile tools and sensors that allow constant iteration of operating plans

Maps provide context for complex public interventions, showing where activities occur in an interconnected system and how the actions of multiple parties may affect outcomes. As the examples in this chapter illustrate, maps can engage the public, create compelling narratives, catalyze collaborations, and inform people's actions to create a healthier and more sustainable environment.

Notes

1. United Nations General Assembly, "Integrated and Coordinated Implementation of and Follow-Up to the Outcomes of the Major United Nations Conferences and Summits in the Economic, Social, and Related Fields," 2005 World Summit Outcome, September 15, 2005, https://www.who.int/hiv/universalaccess2010/worldsummit.pdf.

2. W.M. Adams, "The Future of Sustainability: Re-Thinking Environment and Development in the Twenty-First Century," Report of the IUCN Renowned Thinkers Meeting, January 29–31, 2006, https://portals.iucn.org/library/node/12635.

3. Kirsten Feifel, "PlaNYC: A Comprehensive Sustainability Plan for New York City," Climate Adaptation Knowledge Exchange, April 7, 2010, updated March 3, 2020, https://www.cakex.org/case-studies/planyc-comprehensive-sustainability-plan-new-york-city.

4. City of Seattle, Washington, "Climate Protection Initiative," Government Innovators Network, 2007, https://www.innovations.harvard.edu/climate-protection-initiative.

5. Cathy Wever, "Why going green makes good business sense," *Commercial Real Estate AU*, January 13, 2016, updated April 16, 2019, https://www.commercialrealestate.com.au/news/why-going-green-makes-good-business-sense.

6. National Oceanic and Atmospheric Administration, "What is a watershed," National Ocean Service website, updated February 26, 2021, https://oceanservice.noaa.gov/facts/watershed.html.

7. Mississippi River Cities & Towns Initiative, https://www.mrcti.org.

8. Story in partnership with the US Geological Survey and Geoplatform.gov, 2017, https://geoplatform.maps.arcgis.com/apps/Cascade/index.html?appid=3b88aa4466dc4cb5844ba9ffd394e709.

9. Dale M. Robertson and others, "Spatial Variability in Nutrient Transport by HUC8, State, and Subbasin Based on Mississippi/Atchafalaya River Basin SPARROW Models," *Journal of the American Water Resources Association*, 16 January 2014 https://doi.org/10.1111/jawr.12153.

10. Institute for Public Administration at the University of Delaware and Delaware Department of Transportation, "Sustainable & Resilient—Delaware Flood-Ready Communities," Delaware Complete Communities Toolbox, https://www.arcgis.com/apps/MapJournal/index.html?appid=014f77323c734f3aa03f09d33e44646a.

11. Henry H. Willis and others, "Findings and Policy Considerations," *Current and Future Exposure of Infrastructure in the United States to Natural Hazards*, 2016: 29–32www.jstor.org/stable/10.7249/j.ctt1d9np4s.11.

12. Adopt a Drain, https://adoptadrain.sfwater.org.

13. Stephen Goldsmith and Matt Leger, "A Data-Driven Approach to Cooling a City: Urban heat islands threaten public health. Dallas is turning to a smart growth strategy—and lots of trees—to deal with the problem," *Governing*, December 18, 2018, https://www.governing.com/archive/col-urban-heat-island-public-health-smart-growth.html.

14. Linda Poon, "Cooling Dallas's Concrete Jungle," *CityLab*, August 20, 2018, https://www.citylab.com/environment/2018/08/cooling-dallass-concrete-jungle-plant-1000-trees/566450/?utm_source=feed.

15. Texas Trees Foundation, "Dallas: Urban Heat Island Management Study," August 2017, https://texastrees.blob.core.windows.net/static/Texas%20Trees%20UHI%20Sudy_Final_v4_title%20change.pdf.

16. Cool and Connect Oak Cliff, https://www.texastrees.org/cool-connected-oak-cliff.

17. A Data-Driven Approach to Cooling a City, December 2018, https://www.governing.com/archive/col-urban-heat-island-public-health-smart-growth.html.

18. US Climate Resilience Toolkit, 2019, https://toolkit.climate.gov/case-studies/where-do-we-need-shade-mapping-urban-heat-islands-richmond-virginia.

19. Regional Stormwater Partnership of the Carolinas, "About Stormwater," 2018, https://regionalstormwater.org/about-stormwater.

20. Regional Stormwater Partnership of the Carolinas, "Leveraging GIS Applications to Enhance Municipal Stormwater Management," October 23, 2018, https://regionalstormwater.org/wp-content/uploads/2018/11/Leveraging-GIS-Applications-to-Enhance-Municipal-Stormwater-Management.pdf.

21. US Geological Survey and Geoplatform, “Nutrient Pollution in the Mississippi River Watershed,” 2017, https://geoplatform.maps.arcgis.com/apps/Cascade/index.html?appid=3b88aa4466dc4cb5844ba9ffd394e709.

8

New hope for effective cross-sector collaboration

During the next decade, cross-sector solutions will become the predominant way that societies address their wicked problems. This prediction is premised on three observations:

1. Social problems that challenge our society are a function of multiple, complex, and interlocking issues. Simply addressing one aspect of a problem is rarely enough. Numerous skills and types of interventions are required.

2. Since the 1990s, new types of organizations have defined their mission as the achievement of social outcomes. Many private sector and hybrid organizations have adopted social goals that belonged first to individuals, then volunteer associations, and then governments. Partnerships of all kinds now can address the challenges of wicked problems.

3. Finally, dramatic advances in technology have increased access to data, allowing collaborations to overcome many barriers that once limited

their success. Spatially organized data is not simply a tactical tool; rather, it is a strategic opportunity to create common understanding and collective agency.

This latter development changes the calculus of cross-sector collaboration. Complex problems demand creative solutions that cross policy domains and sectors. Likewise, the emergence of public-private partnerships, consortiums, network alliances, social enterprises, B Corps (companies that meet specific social sustainability and environmental standards), and so on, illustrates the widespread acceptance of multiplayer action. This book has focused on the powerful roles of data and technology in facilitating cross-sector collaboration—a much-needed but difficult-to-manage form of governance.

Overcoming barriers to collaboration

Blending the work of government agencies and community organizations whose operations are often fragmented makes it difficult to coordinate the delivery of complex services. Even as "governing by network"[1] theory advanced cross-sector collaboration in the mid-2000s, integration faced substantial obstacles at the individual, organizational, and community levels. This book has shown, however, that collaborations can use location intelligence to overcome obstacles in these ways:

- **Reach consensus regarding collective objectives.** Individuals and their organizations each start with their own perspectives on what causes a problem, what constitutes an effective solution, and where and how to allocate resources. Sharing well-visualized information and reaching consensus on what it means and implies creates a shared narrative. From that narrative, a collaboration can develop a common set of objectives, agenda, and theory of action.
- **Plan and adapt interventions using real-time rather than static information.** When cross-sector groups form, they commit to a joint endeavor, but each organization has its own goals and culture, and its employees likely have different professional backgrounds. Each partner in the collaboration must understand how its actions affect other partners to avoid wasteful, redundant, inconsistent, or even harmful missteps. Collaborations must understand hyperlocal conditions to allocate resources efficiently, especially in cases of great need and scant

resources. Before the comprehensive digitization of data—gathered continuously and delivered easily—planners necessarily relied on historic information. But even good plans quickly become outdated when partners take actions that change conditions. Concurrently gathered geocoded data allows partners to see impacts (and needs) where and when they occur so that they can adjust their actions in near real time. These dynamic and spatially organized depictions of multiparty action build trust. And applying this jointly acquired knowledge, in turn, improves the capacity for collective action.

- **Act preemptively.** Using data from multiple sources, predictive analytics allows partners to identify trends and anticipate what might happen. Partners can use spatially referenced analytics to assess risk at the neighborhood and block level and, in some cases, even at the building level—for example, in the areas of displacement, crime, and child welfare. Mapped data allows partners to intervene preemptively to prevent problems from occurring or becoming crises. The human and financial ramifications of early intervention significantly improve the efficacy and efficiency of cross-sector partners. An organization may confront an increase in a specific geographically concentrated problem that could get worse and become the responsibility of another partner if not preempted. For example, blood tests can identify areas where a disproportionate number of children suffer from high levels of lead. This GIS-tagged data allows inspectors and other responding partners to redirect resources toward the identification and remediation of the causes. Mapped data of trends and outlier conditions can produce valuable preventive interventions through prediction and preemption.
- **Coordinate the activities of field workers.** Limited sharing of critical and often sensitive information among workers can result in less effective collaboration among partners and their field staff. Before digitization, field workers depended heavily on simple asynchronous messaging. But timely action requires different parties to share and see data from many sources across all actors working in an area. Paper-based processes slow communication, limiting the amount of current information available for timely action. Today's technology dramatically reduces the transaction costs of information sharing. When partners need immediate access to real-time, fully integrated data—such as

during a health, child welfare, or environmental emergency—the ability to share place-based information improves field workers' decision-making. Today's tools allow field workers and their organizations to share privately protected geocoded information enhanced with spatially layered public data with others authorized to see and act on it.

- **Unlock community support.** The success of a collaboration often depends on inspiring confidence, involving more people, attracting more resources, and sometimes changing behaviors. These factors are true whether the workers are volunteers or paid professionals working for a community organization or a city. Building momentum is important too, whether the cause involves protecting a watershed, helping a transitional neighborhood, or reassuring potential marathon runners that comprehensive protection is in place. Success depends on community stakeholders virtually or actually witnessing preparation and progress so that they feel motivated to participate.
  The open data movement allows stakeholders to share progress widely. A park planner and a neighborhood nonprofit can show community members the location of each newly planted tree or how the partnership rebuilt a playground. Neighbors can see new sidewalks and the results of a community trash cleanup effort. Well-visualized action promotes more action, inspiring confidence in people who once felt ignored. High-quality visualization turns open data into inspirational data.

Platforms for modern governance

The changes described in this chapter are only now broadly available. The last decade has seen accelerating breakthroughs in the use of platforms to digitally match buyer and seller, deliver products, and bring people together around common goals. Much of the data these platforms rely on is geographically situated. The same breakthroughs that power Uber and Airbnb also support collective action. In the case of cross-sector collaboration, these platforms facilitate the collection of place-based information from an array of sources, including IoT sensors, outreach workers and residents, and organizational and institutional records. Integrating and mapping data can originate from such sources as field staff who share real-time observations about changing conditions, volunteers who upload GIS and time-stamped photos, IoT sensors that produce constant air or sound monitoring, and residents who call 311 centers or file reports online. The result produces a data-rich picture that collaborations can use to discern need, identify patterns, coordinate operations, preempt crises, and evaluate impacts.

In the public and social sectors, the use of geospatial data, analytics, and mapping is still in its infancy. City government only geocodes a fraction of the information it produces daily. Not enough cities organize their open data around spatial coordinates and thus lack the virtual scaffolding needed for place-based collaboration. Too few nonprofits today have access to GIS tools, and even fewer have adequate training on using location intelligence to support their work. Even when a city government organizes some of its action around a place, it rarely uses that platform to develop more explicit feedback loops and use place-based suggestions from neighbors and residents who want to contribute their thoughts and ideas about community needs. More cities must make it easier to sort open data by geography, enabling comparisons of data across communities. As part of their open data and public improvement efforts, cities must also support and train community leaders to use spatial tools to understand the causes of problems and the actions needed to address them.

These limitations unfortunately restrict the current application of geographic information. Yet experience has shown that the use of analytics and mapping tools supports collaboration, facilitates the exchange of information, and improves coordination among partners across organizations and sectors.

Bridging government and community

Although Americans may disagree on the proper size of government, most agree that government plays a role in public safety, public health, the environment, and the well-being of its citizens. Americans also agree that government has a responsibility to taxpayers and people in need to run as effectively and cost efficiently as possible. Increasingly, Americans understand that government must partner with the social and private sectors to effectively address the complex challenges we face as a society. Place-based thinking and new digital tools can help government and its partners meet these expectations.

In this book, we have showed how collaborations use spatial analytics and visualization to further cross-sector collaboration. We showed how collaborations use these tools in their formative stages to bring potential partners together and understand their stake in working cooperatively. We also explored mapping as a tool to coordinate operations and allocate scarce resources once a collaborative effort is underway. Finally, we showed how collaborations use place-based data to provide insights that allow partners to adapt their interventions in real time.

As we write, national politics are fractious, and action at the federal level is limited. Yet we remain optimistic. In cities across the nation, municipal governments, citizens, nonprofits, and businesses are joining forces to address the wicked problems that challenge our communities. Their combined efforts are strengthened by a shared understanding derived from context—a context provided by the cross-sector use of spatially derived and visualized data.

Notes

1. Stephen Goldsmith and William Eggers, *Governing by Network the New Shape of the Public Sector*, Brookings Institution, 2004, Washington, D.C.

Acknowledgments

We thank the social sector and government leaders who shared their experience with us and who have been tireless in their efforts to improve the lives of the people who live in the communities they serve. We are inspired by their commitment and humbled by their perseverance.

To Kate Bauer and Kate Murphy, we thank you for your ideas, organization, and keen eyes.

To Matt Leger, we owe you a debt of gratitude. Without your insights, diligence, research, and writing, this book would not exist. Thank you.

Study and discussion questions

Introduction

1. Why is the word *wicked* used to define problems such as homelessness, access to health care, and climate change?
    - What makes these problems so complex?
2. What other issues would you classify as wicked?
    - Why do you classify them as such?
    - What constituent issues relate to these wicked problems?
3. What should policy makers consider when addressing wicked problems?
    - Why?
4. If we assume that collaboration is a necessary part of response to complex problems, why do you think multi-stakeholder efforts frequently result in mixed success?
5. Think about a time when you were a member of a team formed to accomplish a specific task.
    - Did you succeed? What factors contributed to your success?
    - What barriers did you have to overcome? How?
    - In what ways, if any, did shared understanding, or the lack of it, contribute to your team's effectiveness?

Chapter 1: Why maps?

1. What is meant by the phrase "what is rooted in systems is experienced in place"?
2. Define cross-sector collaboration.
   - What kinds of cooperative arrangements fall under the umbrella of cross-sector collaboration?
   - Along what dimensions (such as scope, duration, and others) do these cooperative arrangements vary?
3. Think of three collaborative arrangements that fit the definition of cross-sector collaboration. Describe each in terms of its:
   - Goals
   - Partners
   - Scope
   - Duration
   - Degree of formality
4. What are the three phases of collaboration?
   - What conditions and factors contribute to the formation of collaborations, their successful operation, and their ability to adapt?
5. What properties make maps well suited to the demands of cross-sector collaboration? Why?
6. How might the process of negotiating what something means (a piece of text, a set of numbers, geocoded data) lead to shared understanding? In what ways do you think maps facilitate shared understanding?

Chapter 2: Mapping civic engagement

1. Civic engagement takes many forms. Why is place-based or place-specific engagement so vitally important?
2. In what ways can maps be used to encourage civic engagement?

3. Describe the relationship between civic engagement and social capital.
4. What neighborhood or built environment design considerations do urban planners believe encourage the formation of social capital?
5. Stories created using ArcGIS StoryMaps can combine maps, text, imagery, and video content to tell—as their name suggests—a story about an issue or topic. What makes stories a particularly effective tool for civic engagement and advocacy?
6. What is meant by the coproduction of public goods?
7. How does the generation of geotagged input from citizens extend the reach of government?
8. We often think of cross-sector collaboration as occurring between government, nonprofits, and business. But individuals can also be important collaborative partners. What are some of the benefits of incorporating residents' voices and how could maps amplify voice?

Chapter 3: Extending social services

1. Describe a collaboration formed to address a social issue.
   - What conditions or activities helped facilitate or catalyze its formation?
   - What role did individual leaders play in championing its formation?
   - What role did community residents play in championing its formation?
2. Describe a social issue in your community that would benefit from a coordinated response by government, nonprofits, and business.
   - How would you go about engaging potential partners?
     - What data would you use to make the case for collaboration?
     - How would you present that data visually? Spatially?
   - What barriers do you think you might encounter?
     - How might you overcome these barriers?
     - How might you use maps to promote effective collaborative action?

3. The United Way East Ontario (UW) uses location intelligence to support decision-making across its operations. The maps in figures 3.9, 3.10, and 3.11 show how the UW allocates its resources by program type, dollars invested, and people served relative to community socioeconomic status. Using only these maps, what questions would a member of UW's board be justified in asking management?
4. In its work for MCCOY and Northside Ministries, how did SAVI use spatially derived gap analyses to make the case for collaboration and for the reallocation of social service resources?
5. Why did SAVI include church members' home addresses in its geospatial analysis?
   - What does this suggest about the role of personalizing data to engage stakeholders?
   - What potential privacy issues dose this raise?
6. Thrive Chicago and the University of Chicago Urban Labs note that despite the efforts of scores of nonprofits and government agencies, more opportunity youth (young people between the ages of 16 and 24 who are neither enrolled in school nor participating in the labor market) in Chicago become disconnected and at risk of joblessness every year. What is it about how these organizations operate that hinders their effectiveness?
7. How would you apply mapping tools to address the challenges raised by Thrive Chicago and Urban Labs?
   - What data layers would you look at?
   - Where and how might you obtain that data?
8. What do you think about the authors' contention that fragmentation in the delivery of social services is costly?

Chapter 4: Improving public health

1. How would you characterize the information layers captured in figure 4.1?
    - In what ways do they differ?
    - How do they build upon each other?
2. Contrast what happened in Cerro Gordo County, Iowa, and Flint, Michigan. What lessons can we learn from the Iowa example about the application of spatial intelligence to a public health hazard such as arsenic contamination?
3. To identify which of the city's obstetrics and gynecology clinics should immediately respond to the Zika outbreak, the New York City Department of Health and Mental Health assessed risk by neighborhood of pregnant women exposed to the virus. Think about another public health risk that varies by neighborhood. What data might you look at to start planning a response to an emerging problem?
4. Pittsburgh's Breathe Project collaboration involves public health experts, scientists, residents, community groups, and environmental advocates. What digital tools and IoT devices did the collaborative use?
    - How did the use of some of these tools in the hands of residents further the work of the collaborative?
5. What is a satisfaction-adjusted distance measure?
    - How might a group of partners use such a measure to decide where to allocate resources?
6. Consider a public health concern such as prediabetes. If you were forming a collaborative venture in your community to address prediabetes, what types of organizations might you involve?
    - What data would you gather to make the case for collaborating?
    - How would spatial intelligence play into your case?

Chapter 5: Addressing homelessness

1. Why is homelessness considered a wicked problem?
   - What structural, institutional, and individual factors contribute to it?
   - What kinds of geocoded data help policy makers understand the distribution, scope, and dynamics of homelessness in a given community?
2. What strategies have been shown to reduce the incidence of homelessness?
   - In what ways do these strategies presume a cross-sector response to the issue?
   - Why is mapping critical to the implementation of these strategies?
3. Imagine you lead an agency with a specific focus such as domestic violence intervention and prevention or substance abuse intervention or prevention. How would you use geotagged data to make the case for collaboration? To identify potential partners?
4. Describe the dual challenges municipalities face in their street-level responses to homelessness.
   - Why do they require coordination and cooperation?
   - Why are they dependent on location intelligence to coordinate and prioritize their allocation of resources?
5. Healthy Streets San Francisco, the Unified Homelessness Response Center in Los Angeles, and ANCWorks in Anchorage, Alaska, are currently operating collaboratives. What commonalities do you see in how they use geospatial data to facilitate their operations?
6. The Silicon Valley Triage Tool allows outreach workers to prioritize homeless individuals for supportive housing at a net savings to the county. What ethical issues, positive or negative, does the tool's application to the homeless population raise in your mind?

Chapter 6: Responding to disasters

1. During a human-caused or natural disaster, why is it crucial for emergency managers to take a place-based approach to organization and allocation of critical emergency resources?

2. Large-scale disaster response efforts are often fraught with resource allocation and communication challenges. How can geolocation tools and data be used to help minimize or prevent these challenges from impacting the success of response efforts?

3. In coordinating emergency response efforts at the outset of a disaster, how do maps enable a more effective collaboration between multiple stakeholders with competing priorities or information?

4. To help disaster response professionals anticipate storm surge impact by hurricane category, NOAA built the National Storm Surge Hazard Maps. What are the benefits and risks of relying on these predictive models and maps for resource allocation and disaster response?
    - How might emergency managers overcome some of these risks?

5. Mitigating the impact of disasters involves more than public safety professionals and first responders; it also involves private sector partners and everyday citizens. In what ways can these professionals use maps to better involve or produce collaborations with other stakeholders who are not directly responsible for disaster response?

6. In 2015, New York City experienced an outbreak of Legionnaires' Disease caused by infected cooling towers. What was a major impediment for emergency personnel in their ability to respond quickly?
    - What did they do after the outbreak to improve their capabilities for future outbreaks?

7. Why is it important for emergency response professionals to have robust location data about critical stakeholders (for example, nonprofits, shelters, and health-care facilities) in place before an emergency occurs?

    - How can that data be effectively used in the event of a large-scale disaster?
    - Why is it important to consistently stress-test that data and make frequent updates to it over time?

8. Prior to and at the outset of a disaster, how can emergency responders lay the foundation for real-time data sharing and feedback loops that help inform how emergency personnel respond to that disaster?

    - How can they better communicate with first responders as new information comes in that changes tactics or priorities?
    - As first responders enter the field in a disaster, they often encounter unforeseen challenges that the data they had could not possibly predict (for example, a collapsed bridge or roadblocks on key evacuation routes). How might emergency managers account for these unforeseen challenges when they build data feedback loops early on?
    - What challenges do unforeseen events present in accurately displaying data in maps, and how can these challenges be overcome?

9. How can access to maps that display real-time data help to mitigate communication challenges in large-scale disaster response?

10. What are some cross-agency or cross-sector collaboration data sharing opportunities that emergency responders can overlay on maps and use to improve emergency response (for example, using data from the Department of Motor Vehicles to map households without access to a personal vehicle to predict where additional assistance might be needed for evacuation)?

11. How can stakeholders use maps to effectively develop mutual trust during a response, and what effect does the use of maps have on the success of these efforts?

Chapter 7: Increasing sustainability

1. Why do you think the World Summit on Social Development identified economic and social development in addition to environmental protection as two of its three pillars of sustainability?

2. What kinds of collaboration challenges do mayors participating in the Mississippi River Cities Initiative face within and outside their immediate jurisdictions?

   - How does geocoded information help them meet these challenges?

3. Authorizing legislation and a dedicated funding stream are among the antecedent conditions that enable the formation of a collaborative body. The Delaware River Basin is one such body. Why are real-time spatial analytics so crucial to its operations?

4. What is meant by the term *cascading effects*?

   - Why are they salient in watershed management?

5. Adaptive management is a structured approach to decision-making that emphasizes learning and adaptation in response to new information and changing conditions. Why is this approach so well suited to the needs of multiagency or sector partners tasked with watershed management?

6. What are heat islands?

   - How are they related to the wicked problem of health disparities?
   - What role does spatial intelligence play in their mitigation?

7. What have we learned from COVID-19 about the importance of coordination, data, and geospatial intelligence that can be applied to collaboratively addressing climate change?

Chapter 8: New hope for effective cross-sector collaboration

1. In what critical ways do cross-sector approaches differ from conventional approaches to problem solving?
2. What unique challenges do cross-sector and cross-agency collaborations present to municipal governments?
3. What unique challenges do cross-sector and cross-organization collaborations present to nonprofits?
4. In what ways are collaborative partners using spatial intelligence to overcome obstacles to effective coordination and allocation of resources?
5. What do you think needs to happen to make the use of geospatial data, analytics, and visualization more widespread in the public and social sectors?
6. What commonalities did you see among the examples of successful cross-sector or multiagency collaboration presented in this book?
7. Why do the authors believe that cross-sector collaboration will become the dominant approach to addressing wicked problems?
    - Why do you agree or disagree with the authors?
8. How do the principles of wicked problems, cross-sector collaboration, and spatial intelligence apply to your day-to-day work?

Index

A

B

C

D

E

I

K

L

M

N

O

P

R

S

T

About the authors

Stephen Goldsmith was the 46th mayor of Indianapolis and also served as the deputy mayor of New York City for Operations. He is currently the Daniel Paul Professor of Government, director of the Innovations in Government Program, and director of Data-Smart City Solutions at the John F. Kennedy School of Government at Harvard University. He has written *The Power of Social Innovation*; *Governing by Network: The New Shape of the Public Sector*; *Putting Faith in Neighborhoods: Making Cities Work through Grassroots Citizenship*; *The Twenty-First Century City: Resurrecting Urban America*; *The Responsive City: Engaging Communities Through Data-Smart Governance*; and *A New City O/S: The Power of Open, Collaborative, and Distributed Governance.*

Kate Markin Coleman has 30 years of experience as a senior executive in the private and social sectors, where she has successfully participated in multiple significant organizational transformations. As a Harvard ALI fellow, she applied that experience to research that focuses on obstacles to nonprofit efficiency and effectiveness. Kate spent the first half of her career in the private sector. She transitioned to the social sector after she and three colleagues spearheaded the sale of the financial technology company they led. In her subsequent role as executive vice president and chief strategy and advancement officer at YMCA of the USA, Kate led many of the organization's most substantial change initiatives.

About Esri Press

At Esri Press, our mission is to inform, inspire, and teach professionals, students, educators, and the public about GIS by developing print and digital publications. Our goal is to increase the adoption of ArcGIS and to support the vision and brand of Esri. We strive to be the leader in publishing great GIS books and we are dedicated to improving the work and lives of our global community of users, authors, and colleagues.

Acquisitions

Stacy Krieg
Claudia Naber
Alycia Tornetta
Craig Carpenter
Jenefer Shute

Editorial

Carolyn Schatz
Mark Henry
David Oberman

Production

Monica McGregor
Victoria Roberts

Marketing

Mike Livingston
Sasha Gallardo
Beth Bauler

Contributors

Christian Harder
Matt Artz
Keith Mann

Business

Catherine Ortiz
Jon Carter
Jason Childs

For information on Esri Press books and resources, visit our website at esri.com/en-us/esri-press.